Table of Symbols

R	river flow		γ	radioactive half life
S	salinity		Γ	irradiance
$\mathbf{S}$	target strength		δ	specific volume anomaly δ_s, δ_T, $\delta_{S,T,p}$, etc.
t	time			
T	temperature		Δ	difference
T	mass transport or volume transport, period		Δ_{st}	thermosteric anomaly
			ϵ	depletion rate of 0_2
u	speed in x direction		ζ	absorption coefficient of light
U	tangential speed of earth		η	free surface of wave
v	speed in y direction		ϑ	Snell's law angle
$\mathbf{V}$	velocity vector		θ	potential temperature
V	velocity scaler		$\boldsymbol{\theta}$	solid angle
V	volume		κ	wave number
$\mathcal{V}$	group velocity		λ	wave length of light
w	speed in z direction		Λ	wave length of surface wave
W	wind speed		μ	coefficient of molecular viscosity
x	space coordinate, east		ν	scattering coefficient of light
X	a specific distance in x direction		ξ	vorticity
$\mathcal{X}$	unspecified friction		Ξ	residence time
y	space coordinate, north		Π	transmittance
Y	a specific distance in y direction		ρ	density
$\mathcal{Y}$	unspecified friction		σ	density anomaly σ_t, σ_θ
z	space coordinate (parallel to gravity, positive value is ''up'')		Σ	summation
			τ	wind stress
Z	a specific distance in z direction, depth of a layer		Υ	surface tension
			ϕ	latitude, also velocity potential
			Φ	unspecified friction
α	specific volume of water $\alpha_{S,T,p}$, $\alpha_{35,0,p}$, etc.		χ	degree day
			ω	wave frequency
β	attenuation coefficient of light		Ω	angular velocity of earth

Introduction to

PHYSICAL

OCEANOGRAPHY

INTRODUCTION TO

PHYSICAL OCEANOGRAPHY

JOHN A. KNAUSS

University of Rhode Island

Prentice-Hall, Inc. Englewood Cliffs, New Jersey 07632

Library of Congress Cataloging in Publication Data

KNAUSS, JOHN A.
  Introduction to physical oceanography.

  Includes index.
  1. Oceanography. I. Title.
GC150.5.K6   551.4'6   77-14999
ISBN 0-13-493015-0

Printed in the United States of America

10   9   8   7   6   5   4   3   2   1

PRENTICE-HALL INTERNATIONAL, INC., *London*
PRENTICE-HALL OF AUSTRALIA PTY. LIMITED, *Sydney*
PRENTICE-HALL OF CANADA, LTD., *Toronto*
PRENTICE-HALL OF INDIA PRIVATE LIMITED, *New Delhi*
PRENTICE-HALL OF JAPAN, INC., *Tokyo*
PRENTICE-HALL OF SOUTHEAST ASIA PTE. LTD., *Singapore*
WHITEHALL BOOKS LIMITED, *Wellington, New Zealand*

CONTENTS

PREFACE xi

chapter 1

TEMPERATURE, SALINITY, DENSITY, AND OTHER PROPERTIES OF SEAWATER 1

Scale *1*
Temperature *4*
Salinity *8*
Density and the Equation of State *10*
Stability *15*
Some Other Properties of Seawater *18*
Sea Ice *19*

chapter 2

THE TRANSFER OF HEAT ACROSS THE OCEAN SURFACE 28

Sun's Radiative Energy (Q_s) *30*
Back Radiation (Q_b) *33*
Evaporation (Q_e) *34*
Sensible Heat Loss (Q_h) *40*

chapter 3

HEAT, SALT, AND WATER BALANCE 42

Local Heat Balance: The Seasonal Thermocline *43*
Heat Balance and Sea Ice *49*
The Role of Advection in the Global Heat Balance *52*
Global Heat Balance *53*
Salt and Water Balance *58*

chapter 4

CONSERVATION EQUATIONS, DIFFUSION, AND BOX MODELS 61

Conservation Equation *61*
Conservation Equation: Differential Form *63*
Diffusion *65*
Box Models and Mixing Time *68*

chapter 5

EQUATION OF MOTION 72

Acceleration *74*
Pressure Gradient *76*
Coriolis Force *78*
Gravity *84*
Friction *85*
Equations of Motion *89*

chapter 6

BALANCE OF FORCES: WITHOUT CORIOLIS FORCE 91

Hydrostatic Equation *92*
Dynamic Height *95*
River Flow *96*
Convective Flow in an Estuary *97*
Terminal Velocity *106*

chapter 7

BALANCE OF FORCES: WITH CORIOLIS FORCE 111

Geostrophic Flow *111*
Margule's Equation *117*
Wind Stress: Ekman Motion and Upwelling *120*
Inertial Motion *126*
Inclined Plane with Friction *129*
Vorticity and Western Boundary Currents: The Gulf Stream *131*

chapter 8

MAJOR OCEAN CURRENTS 137

Western Boundary Currents: The Gulf Stream and Kuroshio *139*
Currents Along the Eastern Sides of Oceans *150*
Equatorial Currents *151*
Antarctic Circumpolar Current *160*

chapter 9

DEEP CURRENTS AND OTHER OCEAN CIRCULATION 166

Thermohaline Circulation and "Core" Analysis *166*
Formation of Water Types (Masses) *171*
Deep Western Boundary Currents *172*
Characterization of Water Masses and Types *174*
Basins and Sills: The Role of Bottom Topography in Determining
   Temperature and Salinity Distributions *178*
Microstructure *184*
Mesoscale Turbulence: The Role of Eddies *188*
Langmuir Circulation *192*

chapter 10

SURFACE WAVES 194

Wave Characteristics *195*
Particle Motion *199*
Energy and Wave Dispersion *202*
Wave Formation and Capillary Waves *205*
Wave Spectrum and the Fully Developed Sea *208*
Wave Propagation *212*
Refraction and Breakers *215*
Longshore Currents and Rip Currents *217*

chapter 11

TIDES AND OTHER LONG PERIOD WAVES 220

Tsunami *220*
Seiches and Other Trapped Waves *222*
Storm Surges *226*
Internal Waves *227*
Tidal Forces *229*
Equilibrium and Dynamic Theory of Tides *233*
Ocean Tides *235*
Tidal Prediction and Other Changes in Sea Level *238*
Tidal Currents *241*

chapter 12

SOUND AND OPTICS 243

Some Definitions Used in Underwater Sound *244*
Sonar Equation *246*
Refraction and Reflection *253*
Uses of Underwater Sound *258*
Underwater Optics *262*

appendix I

SELECTED DERIVATIONS 269

Conservation Equations: Equations of Continuity *270*
Acceleration *273*
Pressure Gradient *274*
Coriolis Acceleration *276*
Equation of Motion *280*
Friction Shearing Stresses
   (Molecular and Eddy Viscosity) *280*
Reynolds Stress Terms *283*
Diffusion Terms in the Conservation Equation *287*
Vorticity *289*
Small-Amplitude Wave Equation *291*
Group Velocity *295*
Residence Time *296*

appendix II

TABLES OF SOME PHYSICAL PROPERTIES OF SEAWATER *297*

Computing Sigma-*t*
    If One Knows the Temperature and Salinity *298*
Conversion of Sigma-*t* to Thermosteric Anomaly *308*
Conversion of Specific Volume Anomaly to Specific Volume *311*
Conversion of Thermostatic Anomaly
    to Specific Volume Anomaly *312*
Compressibility of Distilled Water and Seawater *316*
Thermal Expansion of Distilled Water and Seawater *319*
Adiabatic Cooling of Water Raised to the Surface *322*
Heat Capacity of Seawater *324*
Molecular Viscosity of Seawater *324*
Velocity of Sound in Seawater *325*
Maximum Density of Seawater, Temperature of Maximum Density,
    and Freezing Point; All As a Function of Salinity *329*
Temperature of Crystallization *330*
Light Absorption of Typical Seawater *330*
Miscellaneous Constants and Conversions *331*

INDEX 333

PREFACE

This text was developed from a one-semester course for beginning graduate students and upper division undergraduates which attempts to introduce the student to the various subjects usually encompassed by the term physical oceanography. The background and interest of the student body vary from marine biologists, for which this may be the student's only formal exposure to the field, to physical oceanographers for which this is only the first of a series of formal courses. Hopefully this text provides sufficient information for non-physical oceanographers to pick their way in the future through that part of the literature which may be of interest to them, and at the same time provides an adequate foundation on which physical oceanographers can build.

The problem faced by the author of any such text is the level of mathematical and physical knowledge to assume, and the reason this text was written was because I found the available texts either too advanced, too elementary, too narrow, or (as in the case of Sverdrup, Johnson and Fleming's, *The Oceans*) a bit out of date. The choice of mathematical level requires a compromise. That chosen here is to assume the reader has some familiarity with those physical principles found in a standard elementary physics text and the mathematics of elementary calculus. Because many students have much more than this background, and because such students usually gain physical insight by looking deeper into the mathematical basis of the stated equations, a number of standard derivations are given in Appendix I.

The second appendix perhaps requires an explanation. Packaged computer programs have apparently all but eliminated the demand for the detailed oceanographic tables that were once on every oceanographer's bookshelf. The hand calculator, however, has made it possible to do quickly a number of calculations at one's desk. Appendix II is an attempt to bring together a variety of tabular material for those who wish to make a few quick calculations that require something more than order of magnitude precision.

This book is affectionately dedicated to that generation of University of Rhode Island students who have lived through and have helped shape a number of earlier mimeographed editions of the notes that eventually became this text. Special acknowledgment is made to several teaching assistants who have helped in catching errors and tracking down references and figures; in particular Philip Richardson, Robert Weisberg, Kenneth Mooney, Paul Temple, and David Lai. Special acknowledgment is also made to Marion Atwood and Anne Barrington who have typed and checked the various versions of notes and texts.

JOHN A. KNAUSS

Introduction to

PHYSICAL
OCEANOGRAPHY

chapter 1

TEMPERATURE, SALINITY, DENSITY, AND OTHER PROPERTIES OF SEAWATER

Scale

A large part of physical oceanography is concerned with the processes responsible for the observed distribution of temperature and salinity in the ocean. Much can be learned about the ocean by observing temperature differences of a few hundredths of a degree, changes in the salt content of a few parts in ten thousand, and changes in density of a few parts in a hundred thousand. This chapter is concerned with some of the processes that account for the observed variations; but before becoming engrossed in the details, it is useful to consider a few typical characteristics of the ocean, as well as a few definitions. The oceans cover about 70% of the earth and have an average depth of something less than 4000 meters (m). Tables 1.1 and 1.2 and Figure 1.1 give some useful figures.

We tend to draw cross sections of oceanic properties with considerable vertical exaggeration (see for example Figure 1.2), but it is well to remember that a typical ocean width is measured in thousands of kilometers and depths are in thousands of meters; thus vertical exagger-

Table 1.1 Area, volume, and mean depth of oceans and sea*

Body		Area (10^6 km^2)	Volume (10^6 km^3)	Mean depth (m)
Atlantic Ocean	⎫	82.441	323.613	3926
Pacific Ocean	⎬ excluding adjacent seas	165.246	707.555	4282
Indian Ocean	⎭	73.443	291.030	3963
All oceans (excluding adjacent seas)		321.130	322.198	4117
Arctic Mediterranean		14.090	16.980	1205
American Mediterranean		4.319	9.573	2216
Mediterranean Sea and Black Sea		2.966	4.238	1429
Asiatic Mediterranean		8.143	9.873	1212
Large Mediterranean seas		29.518	40.664	1378
Baltic Sea		0.422	0.023	55
Hudson Bay		1.232	0.158	128
Red Sea		0.438	0.215	491
Persian Gulf		0.239	0.006	25
Small Mediterranean seas		2.331	0.402	172
All Mediterranean seas		31.849	41.066	1289
North Sea		0.575	0.054	94
English Channel		0.075	0.004	54
Irish Sea		0.103	0.006	60
Gulf of St. Lawrence		0.238	0.030	127
Andaman Sea		0.798	0.694	870
Bering Sea		2.268	3.259	1437
Okhotsk Sea		1.528	1.279	838
Japan Sea		1.008	1.361	1350
East China Sea		1.249	0.235	188
Gulf of California		0.162	0.132	813
Bass Strait		0.075	0.005	70
Marginal seas		8.079	7.059	874
All adjacent seas		39.928	48.125	1205
Atlantic Ocean	⎫	106.463	354.679	3332
Pacific Ocean	⎬ including adjacent seas	179.679	723.699	4028
Indian Ocean	⎭	74.917	291.945	3897
All oceans (including adjacent seas)		361.059	1370.323	3795

*After Kossinna, E., 1921: *Die Tiefen des Weltmeeres*, Berlin University, Institut f. Meereskunde, Veroff., N.F., A. Geogr. naturwiss, Reihe, Heft 9.

Table 1.2 Distribution of depth intervals in world oceans[*]

Ocean	Depth interval in kilometers												Percent of world ocean in each ocean
	0–0.2	0.2–1	1–2	2–3	3–4	4–5	5–6	6–7	7–8	8–9	9–10	10–11	
Pacific Ocean	1.631	2.583	3.250	6.856	21.796	31.987	26.884	1.742	0.188	0.063	0.019	0.001	45.919
Asiatic Mediterranean	51.913	9.255	10.433	12.151	6.698	7.780	1.636	0.076	0.058	0	0	0	2.509
Bering Sea	46.443	5.975	7.623	10.330	29.629	0	0	0	0	0	0	0	0.625
Sea of Okhotsk	26.475	39.479	22.383	3.403	8.260	0	0	0	0	0	0	0	0.384
Yellow and East China Seas	81.305	11.427	5.974	1.239	0.055	0	0	0	0	0	0	0	0.332
Sea of Japan	23.498	15.176	19.646	20.096	21.551	0.033	0	0	0	0	0	0	0.280
Gulf of California	46.705	20.848	25.891	6.556	0	0	0	0	0	0	0	0	0.042
Atlantic Ocean	7.025	5.169	4.295	8.590	19.327	32.452	22.326	0.738	0.067	0.012	0	0	23.909
American Mediterranean	23.443	10.674	13.518	15.313	20.796	13.440	2.572	0.193	0.051	0	0	0	1.203
Mediterranean	20.436	22.475	17.413	30.515	8.940	0.221	0	0	0	0	0	0	0.693
Black Sea	34.965	12.587	23.077	29.371	0	0	0	0	0	0	0	0	0.140
Baltic Sea	99.832	0.168	0	0	0	0	0	0	0	0	0	0	0.105
Indian Ocean	3.570	2.685	3.580	10.029	25.259	36.643	16.991	1.241	0.001	0	0	0	20.282
Red Sea	41.454	43.058	14.920	0.568	0	0	0	0	0	0	0	0	0.125
Persian Gulf	100.000	0	0	0	0	0	0	0	0	0	0	0	0.066
Arctic Ocean	40.673	16.539	10.209	13.167	16.580	2.834	0	0	0	0	0	0	2.620
Arctic Mediterranean	69.013	20.454	6.274	4.260	0	0	0	0	0	0	0	0	0.766
Percent of world ocean in each depth interval	7.192	4.423	4.376	8.497	20.944	31.689	21.201	1.232	0.105	0.032	0.009	0.001	

[*]After Menard, H.W. and S.M. Smith, 1966: "Hypsometry of Ocean Basin Provinces," J. Geophy. Res. 71.

3

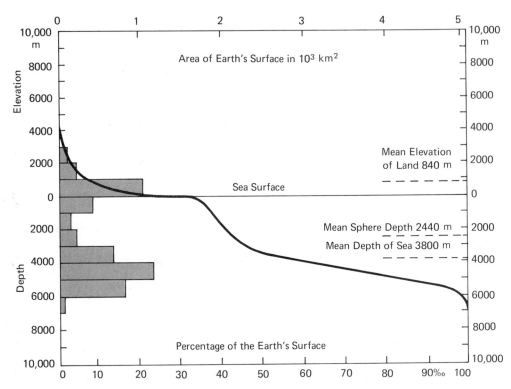

Figure 1.1. Hypsographic curve showing area of the earth's solid surface above any given level of elevation or depth. At left in the figure is the frequency distribution of elevation and depth for 1000-m intervals. (After Sverdrup, H. U., M. W. Johnson and R. H. Fleming, 1942: *The Oceans,* Prentice Hall, Englewood Cliffs, New Jersey.)

ations of 1000:1 or more, as in Figure 1.2, are the rule rather than the exception. Similar exaggeration occurs in thinking about bottom slopes. The "precipitous" plunge down the continental slope from the shelf to the oceanic abyss has a typical slope of 4%.

Temperature

With few exceptions, the temperature of the ocean decreases with depth. Generally, the decrease is more rapid near the surface than at depth. A typical temperature-versus-depth profile has a surface layer tens of meters thick, generally referred to as the *mixed layer*, because surface winds usually play an important role in keeping

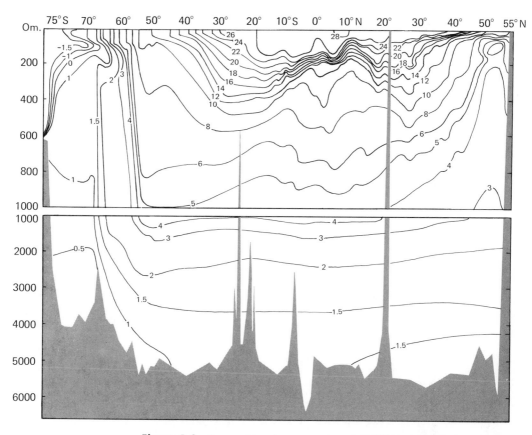

Figure 1.2. Temperature along approximately 160°W in the Pacific from Antarctic to Alaska. Vertical exaggeration is 5.5×10^3 in the upper 1000 m and 1.11×10^3 below 1000 m. (After Reid, J. L., Jr., 1965: "Intermediate Waters of the Pacific Ocean," *The Johns Hopkins Oceanographic Studies,* The Johns Hopkins Press, Baltimore.)

the water well mixed and maintaining a nearly isothermal condition. Below the mixed layer is a region of rapidly changing temperatures referred to as the *thermocline*. The characteristics of the thermocline vary with the season, becoming "stronger" in summer when the mixed layer is warmer, and "weaker" in winter when the surface layer cools. In some ocean areas, such as the western North Atlantic, the *seasonal* thermocline can be distinguished from the deeper *permanent* thermocline; in other areas, including all tropical oceans, the difference between seasons is marked only by a change in the intensification of the permanent thermocline (Figure 1.3).

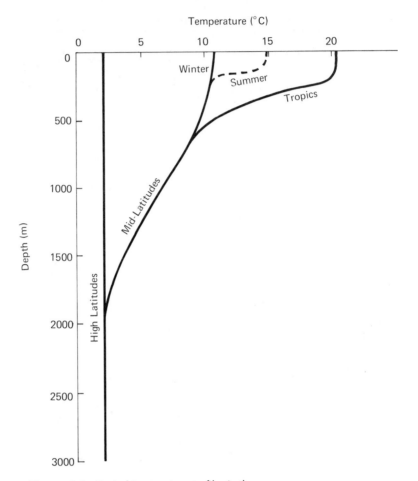

Figure 1.3. Typical temperature profiles in the open ocean.

Below the thermocline the water temperature changes gradually with depth, becoming nearly isothermal again. The water in the lower half of all oceans is uniformly cold. Fifty percent of the water in the ocean is colder than 2.3°C. In 1818, John Ross became the first man to lower a thermometer to any significant depth in the tropical oceans. He found that the water below the surface layer was very cold and drew the obvious conclusion that the deep water must originate at polar latitudes. A typical longitudinal cross section shows a layer of warm water confined to middle and low latitudes, with cold water at depth and at high latitudes (Figure 1.2).

Oceanographers discuss the temperature of the ocean in two

ways, in terms of *in situ temperature* and in terms of *potential temp-erature*. The former is simply the observed temperature. Potential temperature is more complicated. Consider a few liters of water at a depth of 5000 m with an in situ temperature of 1.00°C and a salinity of 34.85 parts per thousand (⁰/oo). Let this water rise to the surface. We shall assume that the rising water is completely insulated from the rest of the ocean water; i.e., there is no exchange of heat or water with its surroundings. The temperature of this water, however, will not be 1.00°C when the water parcel reaches the surface, but about 0.57°C. The potential temperature of the water parcel is 0.57°C.

The concept of potential temperature is easily derived from con-servation of energy considerations (i.e., the first law of ther-modynamics). Stated in words, the change in energy distribution of a mass is

change in internal energy
= heat added or subtracted + work done

If we assume that there is no exchange of heat with the surroundings (i.e., an adiabatic process), the change in internal energy must equal the work done on or by the water. Seawater is slightly compressible, and work is done in compressing the water as it sinks and the water pressure increases. Accordingly, there must be an increase in the internal energy of the water, which means an increase in temperature. The reverse is also true. As the water particle rises, the pressure de-creases, allowing the water to expand. Work is done by the water particle, and the internal energy (thus the temperature) must decrease. A water particle whose temperature is 0.57°C at the surface would have a temperature of 1.00°C if it were brought to 5000 m without mixing or exchanging heat with its surroundings.

In the above example the in situ temperature is 1.00°C, and the potential temperature is 0.57°C. Since the water in the deep ocean ac-quires its temperature characteristics at the surface, there are many instances when it is desirable to know what the temperature of the water would be at the surface (its potential temperature) as well as its observed temperature at depth (its in situ temperature). Table 6 in Appendix II allows one to calculate the potential temperature, know-ing the depth and in situ temperature. The water temperature in all deep trenches, many isolated basins, and in large parts of the North Pacific is isothermal with respect to potential temperature, which

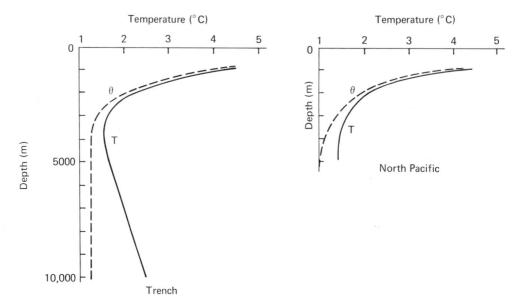

Figure 1.4. In situ and potential temperature curves for a station in the North Pacific at 17°0′N, 162°24′W and in the Mindanao Trench. Sill depth for the Mindanao Trench is about 3500 m, the depth at which the potential temperature curve becomes isothermal.

means that the in situ temperature increases with depth (Figure 1.4). Under such conditions the water is *neutrally stable*, as will be discussed later in this chapter.

The reader may be familiar with adiabatic heating phenomena in the atmosphere, where the effect is much more pronounced. The compressibility of air is much greater than water, and the adiabatic effect in the atmosphere is correspondingly greater. The in situ temperature of a particle of air raised adiabatically decreases between 0.5 and 1°C/100 m depending upon the relative humidity. The warm winds that occasionally sweep down from the mountains, variously called Foehn winds, Santa Anas, Chinooks, etc., get much of their temperature increase by adiabatic heating resulting from the increasing density and pressure of the air as it moves to lower levels.

Salinity

The total amount of dissolved material in seawater is its salinity. Precisely defined, it is ''the total amount of solid materials in grams in

1 kilogram (kg) of seawater when all the carbonate has been converted to oxide, the bromine and iodine replaced by chlorine, and all organic matter completely oxidized.'' The median salinity of the ocean is 34.69 grams (g)/kg of seawater, or 34.69⁰/oo, where ⁰/oo stands for parts per thousand.

No one routinely measures salinity in the manner of the definition. Until about 1955 nearly all salinity measurements were made by determining the amount of the chlorine ion in seawater by titration with silver nitrate, and relating chlorinity to salinity by an empirically derived formula. The presently accepted relationship between the two is

$$\text{salinity} = 1.80655 \times \text{chlorinity} \qquad (1.1)$$

Formula (1.1) carries the explicit assumption that the ratio of dissolved salts in seawater is a constant. To the extent that there are variations in proportions of various dissolved salts (such as variations in the relative proportions of chlorine to calcium or calcium to sulfate), it is not possible to relate salinity to a measure of the halogen content of the seawater. The relationship of Eq. (1.1) holds to about ± 0.02⁰/oo, which is also the standard of precision usually attributed to this method of chemical titration.

Since 1960 most salinity observations have been made by measuring the electrical conductivity of seawater. The latter is a function of temperature as well as salinity; thus a conductivity meter must also measure temperature with great accuracy if it is to provide a measure of salinity. An uncertainty of ±0.01°C in temperature translates approximately to an uncertainty of ±0.01⁰/oo in salinity. Present instruments can provide an accuracy of ±0.003⁰/oo on samples collected and measured in a laboratory. Continuously recording in situ measuring devices are generally less accurate, but recent developments suggest a marked improvement may be possible. The theoretical limit for using electrical conductivity to determine the salt content is also a function of the constancy of the ionic ratios of the dissolved salts. The ultimate precision of electrical conductivity as a measure of salinity is about ±0.001⁰/oo.

Measurements of salinity must be made with considerable precision. For example, the total salinity range of 75% of the ocean is between 34.50 and 35.00⁰/oo, and nearly half the water in the Pacific Ocean is between 34.6 and 34.7⁰/oo. Although there are a large number of problems in the ocean that can be treated adequately, assuming the ocean is of constant salinity, there are also many problems for which changes of only a few hundredths of a part per thousand are important. For example, the slow decrease in salinity in the deep Pacific from about 34.70‰ at 30°S to 34.67‰ at 40°N is evidence for the northward movement of this large volume of water.

It is expected that all the elements found on earth are present in seawater, albeit some are present in such small amounts that they have yet to be detected. Table 1.3 gives the distribution of the most prevalent. Those that are part of the biological cycle, such as calcium, silica, phosphorous, nitrate, and oxygen, vary considerably with depth and location. Also note that, for some purposes at least, the salt content of the ocean can be approximated by sodium chloride, which comprises 85% of the ocean salt.

Density and the Equation of State

The density of seawater is determined by its pressure, temperature, and salinity. The differences in density of the ocean are quite small. Ignoring for a moment the effect on density caused by compressibility, nearly all the water in the ocean has a density between 1.020 and 1.030 g/cubic centimeter (cm³) resulting from changes in temperature and salinity, and the density range of more than 50% of the ocean is only 1.0277 to 1.0279. The largest single effect on density is compressibility. A particle of water with a density of 1.028 at the surface would have a density of 1.051 at 5000 m.

In oceanography, density (ρ) is seldom written out completely. The following convention is used:

$$\sigma = (\rho - 1)10^3 \tag{1.2}$$

Thus for $\rho = 1.02750$, $\sigma = 27.50$.

Whereas oceanographers refer to two types of temperatures, they refer to three types of density:

$$\left. \begin{aligned} \sigma &= (\rho_{S,T,p} - 1)10^3 \\ \sigma_t &= (\rho_{S,T,0} - 1)10^3 \\ \sigma_\theta &= (\rho_{S,\theta,0} - 1)10^3 \end{aligned} \right\} \tag{1.3}$$

Table 1.3 Most abundant elements in seawater by weight.

Element	Concentration (mg/liter)	Element	Concentration (mg/liter)
H	108,000	Pd	
He	0.000005	Ag	0.00004
Li	0.17	Cd	0.00011
Be	0.0000006	In	<0.02
B	4.6	Sn	0.0008
C	28	Sb	0.0005
N	0.5	Te	
O	857,000	I	0.06
F	1.3	Xe	0.0001
Ne	0.0001	Cs	0.0005
Na	10,500	Ba	0.03
Mg	1350	La	1.2×10^{-5}
Al	0.01	Ce	5.2×10^{-6}
Si	3.0	Pr	2.6×10^{-6}
P	0.07	Nd	9.2×10^{-6}
S	885	Pm	
Cl	19,000	Sm	1.7×10^{-6}
Ar	0.6	Eu	4.6×10^{-7}
K	380	Gd	2.4×10^{-6}
Ca	400	Tb	
Sc	0.00004	Dy	2.9×10^{-6}
Ti	0.001	Ho	8.8×10^{-7}
V	0.002	Er	2.4×10^{-6}
Cr	0.00005	Tm	5.2×10^{-7}
Mn	0.002	Yb	2.0×10^{-6}
Fe	0.01	Lu	4.8×10^{-7}
Co	0.0001	Hf	
Ni	0.002	Ta	
Cu	0.003	W	0.0001
Zn	0.01	Re	
Ga	0.00003	Os	
Ge	0.00006	Ir	
As	0.003	Pt	
Se	0.0004	Au	0.000004
Br	65	Hg	0.00003
Kr	0.0003	Tl	<0.00001
Rb	0.12	Pb	0.00003
Sr	8.0	Bi	0.00002
Y	0.0003	Po	
Zr		At	
Nb	0.00001	Rn	0.6×10^{-15}
Mo	0.01	Fr	
Tc		Ra	1.0×10^{-10}
Ru		Ac	
Rh		Th	0.00005
		Pa	2.0×10^{-9}
		U	0.003

Sigma is simply the in situ density, which is a function of salinity, temperature, and pressure [as noted by the subscripts in Eq. (1.3)]. Sigma-*t* is the density that the same particle of water would have if it were at atmospheric pressure (zero gauge pressure). Sigma-θ, potential density, is the density that a particle of water would have if it were raised to atmospheric pressure adiabatically.

In addition, the density of seawater is often referred to in terms of its specific volume, $\alpha = 1/\rho$. As with density, specific volume is seldom written out completely. It is frequently written in terms of the specific volume anomaly (δ):

$$\delta = \alpha_{S,T,p} - \alpha_{35,0,p} \tag{1.4}$$

The specific volume anomaly is the difference between the in situ specific volume and that of seawater at the same pressure, but with a temperature of zero degrees and a salinity of 35⁰/oo.

Table 1.4 and Figure 1.5 were prepared as an aid for keeping the various definitions straight. Note that potential temperature is slightly lower and sigma-θ slightly higher than the corresponding in situ temperature and sigma-*t*. In situ density is seldom used in physical oceanography. Discussion of density is usually in terms of sigma-*t*, sigma-θ, specific volume anomaly, or the specific volume analogy to sigma-*t* (referred to as the *thermosteric anomaly*), or the specific volume analogy to sigma-θ (sometimes referred to as *potential specific volume anomaly*).

Tables for calculating density from known values of temperature, salinity, and pressure are given in Appendix II. Such tables are based on an empirically derived equation of state of seawater. The experimental work on which the equation is based was done prior to 1910. There have been several recent attempts to improve the equation of state. The newer experimental data are of importance in furthering our understanding of the thermodynamics of seawater, but have not yet produced differences that are of significance in studying oceanographic processes. The greatest uncertainty in the equation of state is the effect of pressure on density. There may be uncertainties in the in situ density of several parts in a hundred thousand in the deep ocean, and perhaps 1 part in 10,000 at 10,000 m in the deep trenches. The uncertainty in sigma-*t* (where there is no pressure effect) is less than 0.001, or less than 1 part in 100,000 in density.

Table 1.4 Values of temperature, salinity, density, dynamic height, and stability for a hydrographic station in the North Pacific, 17°04'N, 162°24'W, depth 5726 m

Z	S	T	θ	σ	σ_t	σ_θ	Δ_{st}	δ	$\Sigma\Delta D$	$E(10^{-8}m^{-1})^*$
0	35.003	27.20	27.20	22.68	22.68	22.68	518	518	4.13	—
10	35.000	27.19	27.19	22.73	22.68	22.68	518	519	4.08	—
20	34.997	27.18	27.18	22.77	22.68	22.68	518	519	4.03	—
30	34.995	27.18	27.17	22.81	22.68	22.68	518	519	3.97	—
50	34.992	27.06	27.04	22.93	22.72	22.72	515	517	3.87	2500
75	35.028	25.58	25.56	23.53	23.21	23.21	468	471	3.75	2100
100	35.079	23.83	23.81	24.21	23.77	23.78	414	418	3.64	1800
125	35.096	22.47	22.44	24.72	24.18	24.18	375	380	3.54	1600
150	35.071	21.14	21.11	25.19	24.53	24.54	342	347	3.45	1400
200	34.836	18.10	18.06	26.02	25.14	25.15	283	290	3.29	1200
250	34.438	14.22	14.18	26.85	25.73	25.73	228	235	3.15	1100
300	34.186	10.85	10.81	27.54	26.19	26.20	184	190	3.05	730
400	34.181	8.07	8.02	28.47	26.64	26.65	141	148	2.88	380
500	34.271	6.54	6.49	29.22	26.93	26.93	114	121	2.75	230
600	34.376	5.84	5.79	29.87	27.10	27.11	97	105	2.63	140
700	34.454	5.47	5.41	30.43	27.21	27.22	87	96	2.53	110
800	34.490	4.96	4.90	30.99	27.30	27.30	79	88	2.44	85
1000	34.524	4.14	4.06	32.04	27.42	27.42	67	77	2.28	63
1200	34.552	3.47	3.38	33.05	27.51	27.51	59	68	2.13	48
1500	34.592	2.76	2.65	34.54	27.60	27.61	50	59	1.94	35
2000	34.638	2.07	1.93	36.94	27.70	27.71	41	50	1.67	20
2500	34.663	1.76	1.58	39.25	27.74	27.76	36	46	1.44	12
3000	34.674	1.61	1.38	41.51	27.76	27.78	35	45	1.21	8
3500	34.682	1.52	1.25	43.73	27.78	27.79	33	44	0.99	5
4000	34.688	1.48	1.15	45.91	27.78	27.81	33	45	0.77	3
4500	34.696	1.45	1.06	48.08	27.79	27.82	32	45	0.54	2
5000	34.700	1.45	1.00	50.21	27.80	27.83	31	46	0.32	2
5500	34.700	1.48	0.97	52.31	27.79	27.83	32	48	0.08	—

*E calculated from original data.

Because specific volume is used in calculation of geostrophic currents (see discussion of dynamic height in Chapter 6), most tabulations of density are built around specific volume rather than density. For this reason, and because most of the interesting range of density in the ocean can be expressed in two-digit numbers in specific volume anomaly as compared to three-digit numbers for sigma-*t* and sigma-*θ*, much oceanographic literature is expressed in specific volume anomaly or thermosteric anomaly.

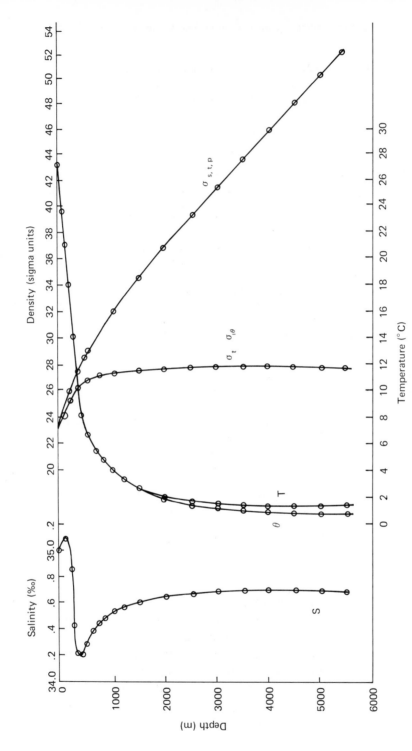

Figure 1.5. Salinity, potential temperature, in situ temperature, σ_θ, σ_t, and $\sigma_{S,T,\rho}$ for a station in the North Pacific at 17°0'N, 162°24'W (see Table 1.4 for numerical values).

The generally accepted nomenclature is as follows:

$$\left.\begin{aligned}
\alpha_{S,T,p} &= \alpha_{35,0,p} + (\delta_s + \delta_T + \delta_{S,T} + \delta_{S,p} + \delta_{T,p} + \delta_{S,T,p}) \\
&= \alpha_{35,0,p} + \delta \\
\alpha_{S,T,0} &= \alpha_{35,0,0} + (\delta_S + \delta_T + \delta_{S,T})
\end{aligned}\right\} \quad (1.5)$$

where the subscripts indicate the functional anomaly. Appendix II includes tables for calculating specific volume as a function of temperature, salinity, and pressure.

The specific-volume counterpart of sigma-t is referred to as thermosteric anomaly.

$$\Delta_{s,t} = \delta_S + \delta_T + \delta_{S,T} \quad (1.6)$$

One can derive the relationship of $\Delta_{s,t}$ to sigma-t, noting that $\alpha_{35,0,0}$ is a constant with a value of 0.97265:

$$\Delta_{s,t} = 0.02735 - \frac{10^{-3}\sigma_t}{1 + 10^{-3}\sigma_t} \quad (1.7)$$

The units of specific volume are volume per unit mass, usually cubic centimeters per gram. A typical value of $\Delta_{s,t}$ would be 0.00155 cm³/g or 155×10^{-5} cm³/g. To keep from writing 10^{-5} when using values of specific volume anomaly or thermosteric anomaly, it has become standard practice to report these values in centiliters per metric tons (cliter/ton). One centiliter per ton is equivalent to 10^{-5} cm³/g.

Stability

Consider first a simple two-layer ocean (Figure 1.6). The work required to move a volume (V) of the denser water (ρ_2) a given distance $Z_1 + Z_2$ upward into the lighter layer (ρ_1) is

$$\text{work} = (\rho_2 - \rho_1)VgZ_1 \quad (1.8)$$

In the absence of friction, no work is required to move the particle of water up to the interface. The work in moving the particle through the

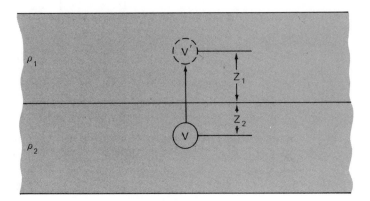

Figure 1.6. In moving a parcel of water (V) of density ρ_2 from its original point to (V'), there is no change in potential energy in moving it the distance Z_2. Work is only required to move it through the layer of density ρ_1.

lighter, less dense layer results in a change of the potential energy of the particle. Often these energy changes are referred to in terms of changes per unit volume (ΔPE). Thus change in potential energy per unit volume is

$$\Delta PE = \frac{\text{work}}{V} = (\rho_2 - \rho_1)gZ_1 \tag{1.9}$$

As the difference in density between the two layers becomes less, the work required becomes less, until the limiting case when $\rho_2 = \rho_1$ and no work is required to move particles of water up and down, and there is no change in the potential energy of the water column.

The density of the ocean changes continuously with depth, and the stability of the ocean can be defined in terms of the rate of change of density with depth:

$$\text{stability } (E) = -\frac{1}{\rho}\frac{\partial \rho}{\partial z} \tag{1.10}$$

(The negative sign is necessary because of the convention adopted throughout this text that the positive z axis points upward.) If the density increases with depth, the ocean is said to be stably stratified. If there is no change in density with depth, and thus no work is required to move a particle from one depth to another, the ocean is said to be neutrally stable. Only rarely, and usually only for brief periods, can one expect to find instances of the density decreasing with depth (static instability). The lighter water would be expected to rise and the heavier water to sink. Most of the ocean is stably stratified, and the stability of the ocean can be used to determine how much work is required to move a particle from one depth to another.

The precise calculation of stability is complicated because water is compressible. The major effect of compressibility will be the tendency for the density of a particle of water to decrease as it is moved upward, since the decreasing pressure allows the water to expand. However, the in situ temperature of the particle will decrease because of adiabatic cooling, which results in a tendency for the density to increase.

For most purposes, stability can be calculated with sufficient precision by noting the change in potential density:

$$E = -\frac{1}{\rho}\frac{\partial \rho_{S,\theta,0}}{\partial z} = -\frac{1}{\rho}\frac{\partial \sigma_\theta}{\partial z} \times 10^{-3} \qquad (1.11)$$

However, because compressibility changes with temperature, even Eq. (1.11) can lead to erroneous conclusions at times. An exact formulation of stability is

$$E = -\frac{1}{\rho}\left(\frac{\partial \rho}{\partial z} - \frac{g}{c^2}\right) \qquad (1.12)$$

where ρ is the in situ density and c is the velocity of sound. The stability values given in Table 1.4 are calculated from the above formula. The reader can make his own calculation of stability using Eqs. (1.11) and (1.12) from the information available in Table 1.4. If he does, he will find no significant differences except in the deep water, where the stability is low.

Oceanographers often use another measure of stability, the Brunt–Väisälä frequency. Imagine a small volume of water at 2000 m enclosed in an insulated balloon so that the enclosed water can expand and contract as necessary as the pressure changes. Lift the balloon of water a few meters above its equilibrium position of 2000 m. Assuming that the ocean is stably stratified at 2000 m, the water in the balloon would be more dense than the surrounding water; if released, it would begin to sink. It would have sufficient speed as it hit the 2000-m mark to pass beyond the point where its density is the same as the surrounding water. It would continue to sink into the deeper heavier water, but its speed would decrease. Eventually, it would stop and then begin to rise, because it would now be lighter than the surrounding water. It would pick up speed until it went through the 2000-m level and again moved into lighter water. In the absence of friction, our imaginary balloon would oscillate back and forth around the 2000-m mark indefinitely. The heavier water from below would push it upward. The lighter water from above would push it back. The larger the density gradient (i.e., the greater the stability), the faster the oscillation. The period of oscillation is given by

$$T = \frac{2\pi}{N}$$

$$N = (gE)^{1/2}$$

(1.13)

where N is called the Brunt–Väisälä frequency.

The shortest periods to be found in the ocean are about a minute (min), corresponding to a stability value of $E \cong 10^{-5}$/cm. In the deep ocean, where the stability is of the order of 10^{-9} to 10^{-10}/cm, the Brunt–Väisälä period is of the order of 3 to 5 hours (h). In those places where the ocean is neutrally stable (regions where the potential temperature gradient is isothermal, see Figure 1.4), the period is infinite.

Care should be taken in the consideration of problems that involve low stability values. Uncertainties in the calculation of density from the equation of state can imply uncertainties in the calculated oceanic stability, which are of the same magnitude as the calculated stability.

Some Other Properties of Seawater

Water molecules are composed of an oxygen atom and two hydrogen atoms bonded as shown schematically in Figure 1.7, which results in a large dipole moment of 1.84×10^{-18} electrostatic units (esu) which in turn is partly responsible for strong interaction of water molecules. Quoting R.A. Horn, "If the water molecules did not have their negative electronic cloud arms and dipole moments they would not be able to interact with one another, the 'oceans' of Earth would be gaseous and life would be impossible."

Any detailed discussion of the thermodynamics or physical

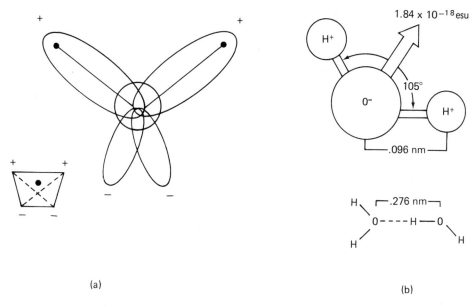

(a)

(b)

Figure 1.7. A schematic view of a water molecule and its bonding.

chemistry of water is well beyond the scope of this text. The structure of liquid water is not well understood; there are a number of models that explain many of the physical properties of water, but not all. Liquid water is anomalous in all its physical–chemical properties (see Table 1.5). Most of the physical–chemical properties of seawater differ in small but significant ways from those of pure water. For typical seawater the velocity of sound is about 4% greater, the compressibility about 13% less, the heat capacity about 5% less, and the freezing point about 2.5°C less. However, in some cases, as for example optical properties, the differences are essentially negligible. The optical transmission properties of filtered seawater are nearly identical with those of pure water. A number of these differences are summarized in Figures 1.8 through 1.13. Tabulated values are given in Appendix II.

Sea Ice

Approximately 3–4% of the ocean surface is covered by sea ice at any one time. Ice plays an important role in the heat budget and circulation of the ocean (see Chapter 3). In this section we limit our

Table 1.5 Some characteristic physical properties of water*

Property	Comparison with other substances	Importance in physical–biological environment
Heat capacity	Highest of all solids and liquids except liquid NH_3	Prevents extreme ranges in temperature Heat transfer by water movements is very large Tends to maintain uniform body temperatures
Latent heat of fusion	Highest except NH_3	Thermostatic effect at freezing point owing to absorption or release of latent heat
Latent heat of evaporation	Highest of all substances	Large latent heat of evaporation extremely important in heat and water transfer of atmosphere
Thermal expansion	Temperature of maximum density decreases with increasing salinity; for pure water it is at 4°C	Freshwater and dilute seawater have their maximum densities at temperatures above the freezing point; this property plays an important part in controlling temperature distribution and vertical circulation in lakes
Surface tension	Highest of all liquids	Important in physiology of the cell Controls certain surface phenomena and drop formation and behavior
Dissolving power	In general dissolves more substances and in greater quantities than any other liquid	Obvious implications in both physical and biological phenomena
Dielectric constant	Pure water has the highest of all liquids	Of utmost importance in behavior of inorganic dissolved substances because of resulting high dissociation
Electrolytic dissociation	Very small	A neutral substance, yet containing both H^+ and OH^- ions
Transparency	Relatively great	Absorption of radiant energy is large in infrared and ultraviolet; in visible portion of energy spectrum there is relatively little selective absorption, hence is "colorless"; characteristic absorption important in physical and biological phenomena
Conduction of heat	Highest of all liquids	Although important on small scale, as in living cells, the molecular processes are far outweighed by eddy conduction

*After Sverdrup, H.U., M.W. Johnson and R.H. Fleming, 1942: *The Oceans*, Prentice-Hall, Englewood Cliffs, New Jersey.

discussion to its physical properties and formation. These can usefully be examined by contrasting sea ice with freshwater ice.

One of the strangest properties of freshwater is the reversal of sign in the thermal expansion coefficient at 4°C. As with nearly all fluids, the density of freshwater increases as the temperature de-

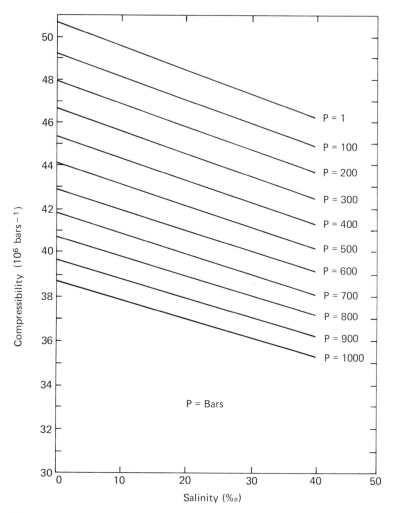

Figure 1.8. Compressibility of seawater at 0°C. It decreases with increasing salinity and pressure (see Appendix II for numerical values).

creases. The rate of increase slows as the water becomes colder, and at 4°C the process begins to reverse; thus water at the 0°C freezing point is lighter than 4°C water. As a typical freshwater lake cools in winter, the colder, denser water sinks until the lake reaches a temperature of 4°C. As the surface water further cools, it floats on the warmer 4°C water, and it is not necessary to cool the entire lake to 0°C to form ice on the surface. The well-known spring "overturn" of northern freshwater lakes results from the warming of the surface

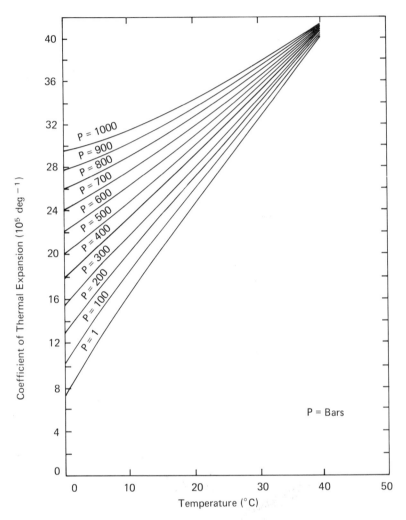

Figure 1.9. Coefficient of thermal expansion of seawater at a salinity of 35⁰/₀₀ of temperature and pressure. It increases with pressure and temperature (see Appendix II for numerical values).

water of these lakes from near freezing to 4°C and the sinking of this warmer water to be replaced by cold, deeper water, which in turn is warmed and then sinks.

As salinity increases, the freezing point decreases; the inflection point of the thermal expansion in turn moves to lower temperatures (Figure 1.11 and Table 10, Appendix II). At a salinity of 18.56⁰/₀₀, the

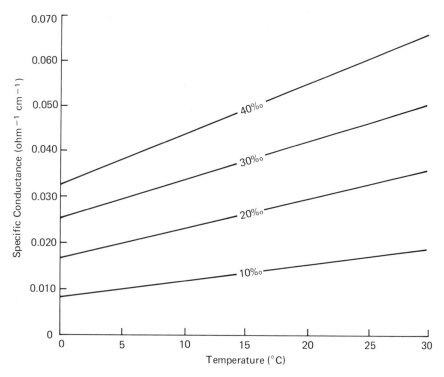

Figure 1.10. Effect of temperature on the specific electrical conductivity of sea-water of various salinities at 1 atm. It increases with increasing temperature and salinity.

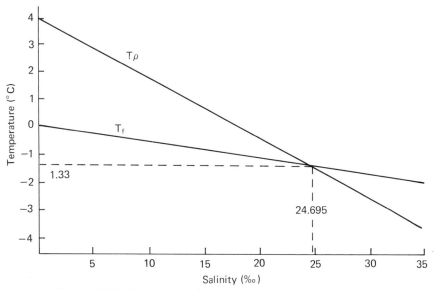

Figure 1.11. Temperature of density maximum $T\rho$ and temperature of freezing point T_f for seawater of different salinities (see Appendix II for numerical values).

23

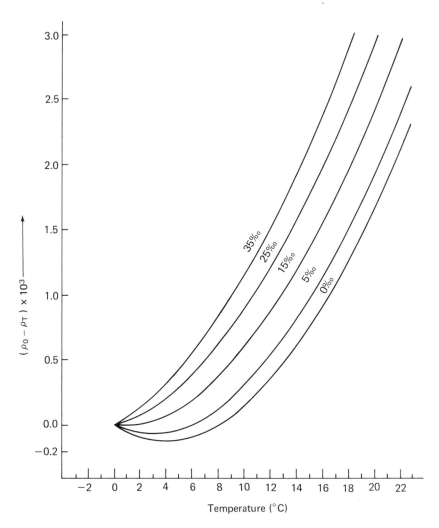

Figure 1.12. Density of pure water and of seawater of different salinities as a function of temperature. ρ_0 is the density of 0°C, and ρ_T is the density of T°C (for detailed tabular values see Appendix II).

inflection point is 0°C; at a salinity of 24.7⁰/oo, the inflection point and the freezing point are −1.33°C.

The thermal expansion properties of seawater with a salinity greater than 24.7⁰/oo behave in a more orthodox fashion; that is, the density of seawater continues to increase as the temperature decreases until the freezing point is reached. A logical consequence of

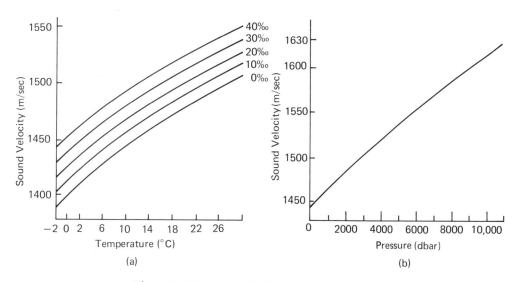

Figure 1.13. (a) Sound velocity in seawater as a function of salinity and temperature at atmospheric pressure; (b) sound velocity in seawater at 0°C and 35⁰/₀₀ as a
function of hydrostatic pressure (see Appendix II for numerical values).

this effect would be the requirement that the entire ocean basin be
cooled in winter to the freezing point before sea ice could form.
Otherwise, the colder water would sink. However, an inspection of
Arctic temperatures indicates no ocean basins with the temperature
of −2.5°C required for freezing water of 34⁰/₀₀. What one finds instead
in areas of ice formation is a sharp *halocline* at a depth of 50–200 m. A
halocline is a layer of rapidly changing salinity, analogous to a thermocline. The fresher, and therefore lighter, layer of surface water can
be cooled to the freezing point without becoming denser, and therefore unstable with respect to the deeper saltier water below (Figure
1.14).

At some risk of oversimplification, the formation of sea ice can
be considered simply as the freezing of the freshwater, leaving the
salt behind in brine pockets. As the temperature reaches the freezing
point, ice crystals of pure water are formed, which "surround" the
unfrozen water. This unfrozen water is enriched in salt left behind by
the frozen crystals, which results in a further lowering of the freezing
point of these brine pockets. The ice crystals are plate shaped, often
hexagonal, with horizontal dimension of about 2.5 cm and a vertical
dimension of about 0.5 millimeters (mm). Since sea ice is a much

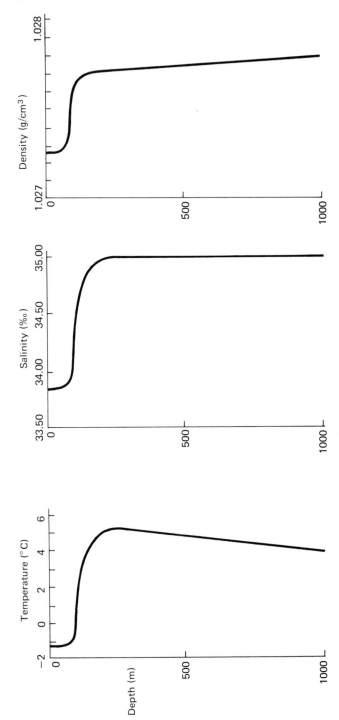

Figure 1.14. Typical salinity and density profiles in areas of ice formation. Note the temperature inversion and the very sharp pycnoline, whose shape is primarily determined controlled by the halocline.

better thermal conductor than brine, new ice crystals are generally formed by aggregating on those crystals already present, rather than forming independently in the brine. Ice crystals tend to form sheets, and there is evidence to suggest that those sheets which are tilted toward the vertical grow faster than those which are horizontal. If the frozen crystals do not completely surround the salt-enriched unfrozen water, this water will sink and mix with the seawater below. If the freezing is slow enough, nearly all the enriched brine will escape, and the salinity of the sea ice will be approximately zero. Rapid freezing entraps most of the brine, resulting in a sea ice salinity close to that of the surrounding water. Most sea ice salinities range from 2–20⁰/oo, with the older ice averaging lower salinities since leaching of brine in old ice is enhanced by alternate melting and freezing as the air temperature changes. If the temperature is low enough, the salt itself will begin to crystalize. Sodium sulfate begins to crystalize at $-8.2°C$ and the important sodium chloride at $-23°C$ (Table 11, Appendix II).

Because sea ice has brine pockets and complex salt crystalization patterns, it is less easy to characterize its strength and similar physical attributes than is the case for freshwater ice. Generally, sea ice has about one-third the strength of freshwater ice of the same thickness. However, it may be comparable for old ice (with very low salinities) and for sea ice at a temperature well below the crystalization temperature of sodium chloride.

chapter 2

THE TRANSFER OF HEAT
ACROSS THE OCEAN SURFACE

For most purposes we can assume that all the heat enters the ocean at the sea surface. The only other significant source of heat is the earth itself; however, $\frac{1}{10}$ calorie (cal)/cm²/day that reaches the ocean through the floor of the ocean is small compared to the average value of 400 cal/cm² of the sun's radiative energy that is absorbed by the surface layer of the ocean each day.

To a high degree of approximation, the temperature of the ocean has not changed in recent times. Comparisons of the first adequate deep ocean temperature measurements made more than 60 years ago with those made today indicate that any changes that may have taken place in this period are less than our ability to observe; thus we may further assume that the amount of heat entering the ocean is equal to that leaving the surface. Significant heat is exchanged across the ocean surface by four processes: (1) shortwave radiation is received from the sun; (2) radiative heat in the infrared spectrum is exchanged between the atmosphere and the ocean; (3) heat is lost from the ocean as water is evaporated; and (4) heat is exchanged between the atmosphere and the ocean whenever there is a temperature difference between the sea surface and the atmosphere.

One can write a simple balanced equation that equates the rate at which heat enters the surface of the ocean to the rate at which it leaves:

$$Q_s = Q_b + Q_e + Q_h \tag{2.1}$$

where Q represents the flux of heat energy across a unit area of sea surface and the subscripts s, b, e, and h refer, respectively, to the incoming radiation of the sun, the net long-wave energy radiated back by the ocean, the heat loss by evaporation, and the sensible heat loss (or gain) resulting from heat conduction across the air–ocean interface.

Although there is an overall oceanic heat balance, these four terms do not need to balance locally. For example, heat is added to the surface layer of the ocean during the summer and raises the temperature of the water, and the equivalent amount of heat is lost during the winter as the surface layer cools. Nor is it necessary that the four terms of Eq. (2.1) balance exactly when averaged over a year. The ocean receives more heat energy in tropical latitudes than it loses. Warm water is transported northward, and in polar regions the ocean gives up more heat across the sea surface than it receives. The ultimate cause of both atmospheric and ocean circulation is due to the fact that there is a net gain (loss) of heat energy at low (high) latitudes.

To allow for the storage or release of heat in the surface layer of the ocean, we can write

$$Q_T = Q_s - Q_b - Q_e - Q_h \tag{2.2}$$

where Q_T represents the rate at which heat is added to or lost from the surface layer.

Although we know there is a heat balance in the ocean, our detailed knowledge of how this balance is maintained is quite unsatisfactory. The processes we understand best relate to the sun's radiative heat energy entering the ocean. There is some disagreement on the relative importance of the other heat-balance terms, and there is considerable uncertainty on the details of the processes by which heat is exchanged, as well as the numerical values to be assigned.

In this chapter we discuss the individual terms of Eq. (2.1). In Chapter 3 we take up the question of heat storage and loss in the surface layer.

Sun's Radiative Energy (Q_s)

A flat plate beyond the earth's atmosphere placed perpendicular to the rays of the sun receives about 2 cal of heat/cm²/min. The exact figure will vary a few percent with the distance of the sun from the earth and with sun spots and solar flares. The average value is called the *solar constant*. Its currently accepted value is between 1.92 and 1.96 langleys/min (1 cal/cm² is 1 langley, ly). There is negligible energy loss in space, and the energy spectrum measured outside the atmosphere approximates that of a black-body radiation source with a temperature of about 6000°C (Figure 2.1). Some 49% of the energy is in the visible spectrum with wavelengths between 400 and 700 nanometers (nm) (1 nm is 10^{-9} m or 10^{-7} cm), 9% in the ultraviolet, and 42% in the infrared. The maximum energy is at about 500 nm. Some 99% of the sun's energy has a wavelength less than 4000 nm.

At any given moment the earth is receiving an amount of energy equal to the solar constant times the cross-sectional area of the earth, πR^2, where R is the earth's radius. This energy is on the average distributed over the surface of the earth, $4\pi R^2$, so that the average amount received is about 0.5 ly/min or 700 ly/day. At any given position on the earth the amount varies with the declination of the sun. At the poles it can vary from 0–1100 ly/day and at 42° latitude from about 300–900 ly/day. The amount of radiation that would reach the earth, assuming no clouds or atmospheric absorption, is given in Table 2.1.

Once the sun's rays enter the earth's atmosphere the energy is both scattered and absorbed. Both processes are selective; for example, much of the ultraviolet energy is absorbed in the upper atmosphere in various photochemical reactions, including the production of ozone. On the other hand, more than 75% of the energy in the visible spectrum reaches the surface on clear days. However, a significant portion is scattered by the molecules in the earth's atmosphere and, as a result, reaches the earth as diffuse sunlight. If the sun's radiation were not scattered, the sky would be black, as it appears to as-

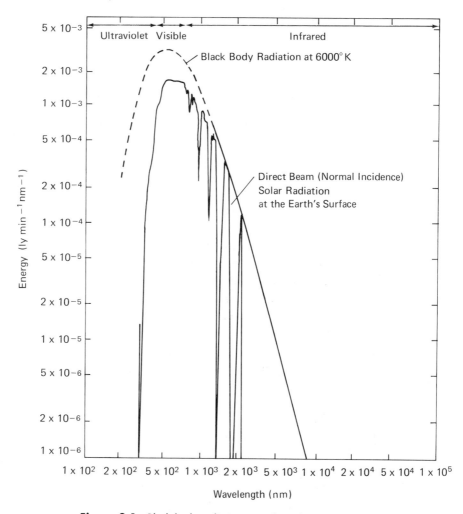

Figure 2.1. Black-body radiation curve from the sun and observed values at the sea surface on a very clear day. Nearly all the ultraviolet and all the infrared above 2200 nm are absorbed before the radiation reaches the surface. (After Sellers, W. D., 1965: *Physical Climatology,* University of Chicago Press, Chicago.)

tronauts circling in outer space. The blue of the sky is because the molecular scattering increases as the inverse fourth power of the wavelength:

$$\text{scattering} \cong \frac{1}{\lambda^4} \tag{2.3}$$

Table 2.1 Amount of solar radiation incident upon the surface of the earth under a completely transparent atmosphere (values in kcal cm^{-2})

	Latitude									
	0°	10°	20°	30°	40°	50°	60°	70°	80°	90°
Summer half-year	160.5	170	175	174	170	161	149	139	135	133
Winter half-year	160.5	147	129	108	84	59	34	13	3	0
Year	321	317	304	282	254	220	183	152	138	133

Thus blue light (λ = 400 nm) is scattered 10 times more effectively than red light (λ = 700 nm). This Rayleigh scattering law only holds when the particles doing the scattering are small compared to the wavelength of the light. When the particles are of the same size or larger than the wavelength, the scattering is less dependent upon the wavelength. A person looking skyward and not directly at the sun perceives primarily scattered sunlight. A blue sky is a consequence of Rayleigh scattering; a deeper blue means simply that a greater percentage of the skylight is a result of Rayleigh scattering. Aerosols, dust, and water particles cause more uniform scattering and a grayer, less blue sky.

Rayleigh scattering also occurs in the ocean; the deep blue of open ocean, as contrasted with the green or gray of coastal waters, is due in part to the selective scattering of the shorter wavelength.

Difference in cloud cover causes the greatest variation in the amount of radiation reaching the surface. A dense cover of low stratus clouds can absorb or reflect back to space more than 80% of the radiation from the sun. An average value of about 25% has been estimated for the amount of incoming energy of the sun that is cut off by clouds.

The average amount of energy reaching a point on the earth will vary with season, latitude, and cloud cover. Charts drawn by Budyko show values ranging from 60 × 10^3 ly/year to more than 200 × 10^3 ly/year, with the higher values in the deserts. The highest value for an oceanic area is about 160 × 10^3 ly/year, which is equivalent to 0.30 ly/min. In other words, half the radiative energy of the sun reaches the surface of the ocean as either direct or scattered sunlight. Of this amount, something less than 10% is reflected back from the ocean surface.

The *albedo* of an object is the percent of radiation reflected from its surface; it varies somewhat with the angle the sun makes with the reflecting surface. If the wind is blowing hard and the sea surface is rough, the angular dependence is less than otherwise. At high sun angles the reflectance may be only 3%, but at very low sun angles it may be as high as 30%. Commonly accepted "average values" of the sea surface albedo center around 6%. By way of comparison, the albedo for sea ice is approximately 30–40%, and for fresh snow it can be as high as 90%.

Back Radiation (Q$_b$)

The ocean also loses heat through radiation. All bodies with a temperature above absolute zero radiate heat energy. The amount is proportional to the fourth power of the absolute temperature (Stefan–Boltzmann law):

$$Q = cK^4 \tag{2.4}$$

where K is the absolute or Kelvin temperature, and c, the Stefan–Boltzmann constant, has a value of 1.36×10^{-12} cal cm^{-2} second^{-1} (s^{-1}) K^{-4}. The wavelength of maximum transmission is inversely proportional to the absolute temperature (Wien's law):

$$\lambda_{max} = \frac{c'}{K} \tag{2.5}$$

where $c' = 2.9 \times 10^{+6} K$ nm. Thus the radiation power of a square centimeter of the sun's surface with a temperature of 6000°C is about 200,000 times that of a square centimeter of sea surface at a temperature of 10°C. While the maximum energy of the sun's radiation is 500 nm, that of the sea surface is about 10,000 nm.

It is easy to calculate the theoretical radiation loss of the sea surface from the Stefan–Boltzmann law. If this is done, it would appear that the sea surface loses more energy than it is receiving from the sun. For example, a Kelvin temperature of about 233°K (or 40° below zero) is required to radiate the equivalent of 4×10^{-3} ly/s. The latter is the Budyko figure for the average amount of incoming solar radiation. For an ocean with an average surface temperature of about 17°C, the equivalent black-body radiation is well over twice that amount.

The solution to this apparent paradox is that much of the long-wave radiation from the sea surface is absorbed by the clouds and water vapor in the air and *reradiated* back to the surface. The useful term is not the back radiation as calculated by the Stefan–Boltzmann law but effective back radiation. This is defined as the net long-wave radiation loss from the sea surface. This term varies with the water vapor in the air, a fact that can be at least qualitatively verified by noting how much cooler it can become on a clear dry night than on a cloudy one. On clear evenings most of the long-wave radiation escapes into space. On a cloudy night, or one with a high relative humidity, much of the long-wave radiation is absorbed by water vapor and reradiated back to the earth's surface. Various investigators have estimated that the effective back radiation from the ocean is about 100–125 ly/day. All agree that the controlling factor is not the sea surface temperature but the water-vapor content in the atmosphere.

A comparison of the incoming radiation of the sun with that of net long-wave radiation from the surface of the earth shows a net heat gain to the earth. The amount of this radiation balance is much larger over the ocean than land and, as a comparison of Figures 2.2 and 2.3 will show, less than half the incoming radiation is accounted for by the net long-wave back radiation of the ocean surface.

Evaporation (Q_e)

The amount of heat required to evaporate 1 g of seawater varies slightly with both temperature and salinity. An average value is 590 cal/g. It has been estimated by Budyko that the oceans lose the equivalent of a 126-cm layer of water/year through evaporation. In terms of energy loss, this is 200 ly/day. This estimate is admittedly a difficult one to make, and others have calculated somewhat lower figures. However, all such estimates result in the conclusion that heat loss by evaporation is the largest of the three heat-loss terms in our simple balance equation, Eq. (2.1). Our inability to measure this term in the ocean is a source of continuing frustration. The statement made by Mathew Fontaine Maury more than a hundred years ago continues to

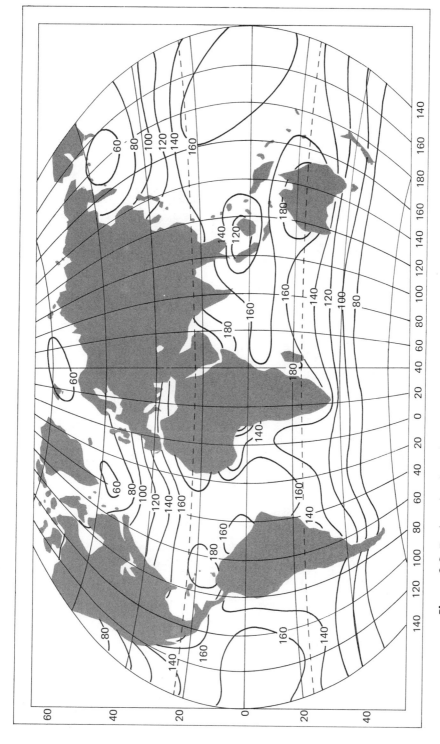

Figure 2.2 Total yearly solar radiation received at the surface of the earth (values in kcal cm^{-2} yr^{-1}). (After Budyko, M. I., 1974: *Climate and Life*, Academic Press, New York.)

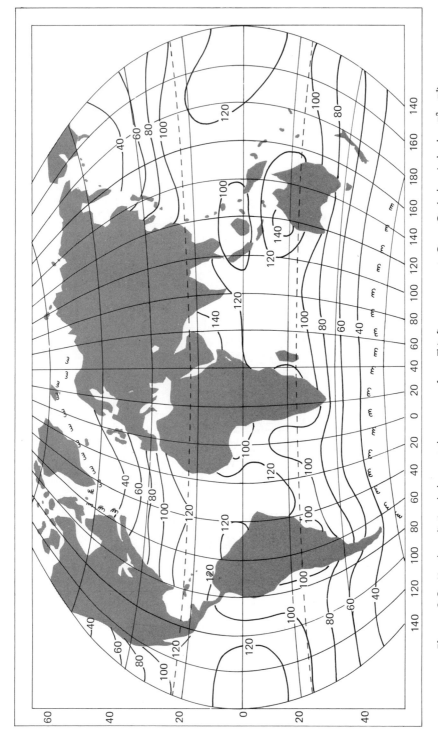

Figure 2.3. Yearly radiation balance at the ocean surface. This figure represents $Q_s - Q_b$ (values in kcal cm^{-2} yr^{-1}). (After Budyko, M. I., 1974: *Climate and Life*, Academic Press, New York.)

have some validity: "The state of our knowledge of evaporation that is daily going on at sea has notwithstanding the activity in the fields of physical research been but little improved."

The nature of the process and the measurement problem can be appreciated by considering a simple, nonturbulent model. Imagine a stationary, isothermal layer of air overlying a quiet layer of water of the same temperature as the air; in other words, a situation where there is no turbulence and where there is no warmer underlying water to heat the air and convect it upward. In this case one could make a reasonably accurate calculation of the rate at which water is evaporating, if one knows only the temperature of the water and the relative humidity of the air. Furthermore, one could verify these calculations by laboratory experiment. The ocean situation, however, is very different. The wind blows, resulting in waves on the surface and turbulence in the air column above. If the wind blows hard enough, the waves break and spray is thrown into the air, which enhances evaporation. Even in the absence of wind, the ocean surface is usually a degree or two warmer than the air above, which means that the bottommost, moisture-laden layer of air is warmed by the ocean and convected upward; it is replaced by drier air, which in turn becomes water saturated as evaporation continues on the molecular level at the air–sea interface.

The highly turbulent nature of the flux regime makes meaningful observations very difficult, and various theoretical attempts require either empirical constants of large uncertainty or difficult-to-achieve observational data on fluctuations of properties and gradients. A number of attempts have been made to calculate evaporation using different approaches, and often quite different assumptions. Generally, these different methods of calculation give answers that agree with one another to within a factor of 2, and often they are much more consistent than that.

No one has yet succeeded in making successful evaporation measurements at sea. One can measure the water evaporated from a pan on the deck of a ship or an offshore tower, but this can only crudely approximate the true evaporation. Others have tried to estimate the rainfall over the ocean and the amount of water running into oceans from rivers. The total water evaporated must approximate the sum of the two; otherwise, sea level would be changing.

Charts indicating annual and seasonal evaporation rates of the oceans have been published (Figure 2.4). All such charts are based on equations that are checked with empirical evidence where possible. The equations have some theoretical justification, but the authors themselves would be the first to admit that the numbers as well as the equations are at best an approximation. All assume that the evaporation rate is controlled by turbulent processes rather than molecular processes, and that evaporation should increase with increasing wind speed, which increases the turbulent transfer processes. Evaporation should also increase if the surface layer of air is unstable; that is, if the water surface is warmer than the air, causing small convection processes to occur, which remove the boundary layer of air saturated with water vapor and allow it to be replaced with unsaturated air, which can, in turn, accept more water vapor through evaporation. An extension of the same argument suggests that evaporation increases when the relative humidity of the overlying air is low, since the capacity of the air for accepting water vapor is proportionately higher.

Some of the highest evaporation rates in the ocean are estimated to occur in winter near the Gulf Stream, where cold dry air from Canada is brought down over the warmer ocean water. Jacobs estimated the winter evaporation rate for this region might be as high as 1 cm/day (600 ly/day), a figure which is not inconsistent with the more recent work of Budyko. Jacobs's estimates of evaporation were based on the simplest of empirical formulas:

$$Q_e = c(e_w - e_a)W \tag{2.6}$$

where e_a is the vapor pressure in the air some distance above the water (deck height aboard a ship), which can be found from knowing the air temperature and the relative humidity, and e_w is the vapor pressure of the air at the water surface (assuming that the air temperature is the same as the water temperature, and the relative humidity is 100%). The turbulence factor is parameterized by using the wind speed W, and c is a numerical constant. For a wind speed in knots, vapor pressure in inches of mercury, and heat loss in langleys per day, the value of c used by Jacobs was 145.

A different technique has been used by other investigators, including Budyko, based upon the concept that similar processes control evaporation and sensible heat loss by convection. It is the Bowen ratio and is discussed in the next section.

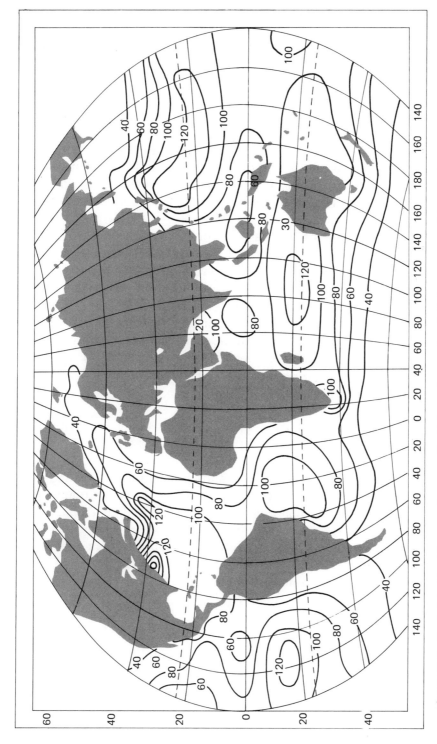

Figure 2.4. Yearly latent heat flux over the ocean. The highest values are found off the east coast of the United States over the Gulf Stream. (After Budyko, M. I., 1974: *Climate and Life*, Academic Press, New York.)

Sensible Heat Loss (Q_h)

The third important way in which heat is lost from the surface of the ocean is by convection and conduction. On the average, the surface of the ocean is warmer than the air. Thus the ocean can be expected to give up heat to the atmosphere by the process of heat conduction. Again, however, the heat loss is not controlled by molecular processes. If it were, the rate of heat transfer would be simply proportional to the air–sea temperature differential. Empirical formulation of the heat-transfer equation indicates that heat transfer increases with increasing wind speed, where wind speed is used as an index of the rate of turbulent transfer across the ocean–atmosphere interface.

The problems of measuring or calculating sensible heat transfer are similar to those described for evaporation; however, the uncertainty in our theoretical and observational base is not so critical, since the estimates of sensible heat transfer are of the order of 10–15% of that calculated for evaporation.

Since the density stratification and the turbulent processes at the air–sea boundary affect both the evaporation and convection of sensible heat, a number of attempts have been made to relate the two, beginning with Bowen in 1926. In its simplest form, Bowen's ratio B can be written as

$$B = \frac{Q_h}{Q_e} = 0.5 \frac{T_w - T_a}{e_w - e_a} \tag{2.7}$$

where $e_w - e_a$ is the vapor pressure gradient in millibars as defined in the previous section, and $T_w - T_a$ is the analogous temperature gradient. The numerical constant 0.5 assumes that identical turbulent processes determine both Q_h and Q_e; more precisely, the numerical values for vertical eddy diffusion coefficients for heat and water vapor are identical (see Chapter 4). A number of attempts have been made to calculate evaporation by combining Eqs. (2.7) and (2.2). Knowing Q_e, one can then calculate Q_h:

$$Q_e = \frac{Q_s - Q_b - Q_T}{1 + 0.5 \left(\dfrac{T_w - T_a}{e_w - e_a} \right)} \tag{2.8}$$

Over most of the ocean, the air–sea temperature differential is less than 2°C. The exceptions are near continents and in ice fields. The annual oceanic range of Q_h suggested by Budyko is from more than 130 ly/day off the East Coast of the United States to a -50 ly/day (heat transferred from the air to the ocean) in the South Atlantic south of 40°S, where ice cover becomes an important factor. The average value for the ocean is about 25 ly/day.

chapter 3

HEAT, SALT, AND WATER BALANCE

In their simplest form the conservation equations are so self-evident that often they are not explicitly stated. For example, the amount of heat entering the ocean must equal the heat leaving the ocean or the average temperature of the ocean will change. The amount of water entering the ocean by rainfall and river runoff must equal the amount lost by evaporation or sea level will change. The amount of salt brought into the oceans by rivers must equal the amount removed by precipitation and other means or salinity will change. In other words, in a steady-state situation, what goes in must equal what goes out. If it does not, then there is a change in temperature, volume, or salinity, and this rate of change can be calculated assuming the necessary parameters are known. Thus it is possible to calculate the seasonal change in the temperature of the surface layer of the ocean by knowing the difference in the heat coming in and going out and the specific heat and density of the water.

In Chapter 4 the conservation equations will be treated in a more formal manner. The succeeding discussion can be adequately understood on the basis of either rate of change = what goes in − what goes out

or, in the case of a steady state,

what goes in = what goes out

In Chapter 2 we discussed the terms responsible for the flux of heat across the air–water interface:

$$Q_s = Q_b + Q_e + Q_h \tag{3.1}$$

Strictly speaking, this equality exists only if one averages over the entire ocean surface (s) and at least one year (t):

$$\int^t \int^s (Q_s - Q_b - Q_e - Q_h) \, ds \, dt = 0 \tag{3.2}$$

Heat is transported poleward by major currents such as the Gulf Stream; as a result, the ocean loses more heat at high latitudes than it receives from the sun. Similarly, the surface layer of the ocean stores heat in summer and releases it in winter. Thus the nonintegral form of the heat-balance equation requires two additional terms:

$$Q_s = Q_b + Q_e + Q_h + Q_T + Q_V \tag{3.3}$$

where Q_T refers to the heat used for warming the ocean, and Q_V refers to the heat transported into or out of the area.

Local Heat Balance: The Seasonal Thermocline

Once the sun's radiation penetrates the surface of the sea, it is rapidly absorbed. The ocean is nearly opaque to all forms of electromagnetic radiation, from the very long radio waves to the very short ultraviolet. What small transmission window exists is in the visible spectrum, but even here the transmission is extremely poor compared to the atmosphere. In even the clearest ocean waters, less than 1% of the energy reaches 100 m, and all but the visible portion of the spectrum is absorbed in the first few centimeters. Table 3.1 gives typical values for the percent of energy reaching a given depth for a mid-ocean area with maximum transmission and a coastal estuary

Table 3.1 Percentage of total irradiance (300–2500 nm) that reaches
a given depth in typical oceanic and coastal waters*

Depth	Oceanic water†					Coastal water				
(m)	I	IA	IB	II	III	1	3	5	7	9
0	100	100	100	100	100	100	100	100	100	100
1	44.5	44.1	42.9	42.0	39.4	36.9	33.0	27.8	22.6	17.6
2	38.5	37.9	36.0	34.7	30.3	27.1	22.5	16.4	11.3	7.5
5	30.2	29.0	25.8	23.4	16.8	14.2	9.3	4.6	2.1	1.0
10	22.2	20.8	16.9	14.2	7.6	5.9	2.7	0.69	0.17	0.052
20						1.3	0.29	0.020		
25	13.2	11.1	7.7	4.2	0.97					
50	5.3	3.3	1.8	0.70	0.041	0.022				
75	1.68	0.95	0.42	0.124	0.0018					
100	0.53	0.28	0.10	0.0228						
150	0.056			0.00080						
200	0.0062									

*From Jerlov, N.G., 1968: *Optical Oceanography*, Elsevier Publishing Company, New York.
†For oceanic the solar altitude is 90°; for coastal water, 45°.

area, where the turbid water results in even more rapid absorption of
the light energy. Note that in either case more than 50% of the sun's
energy is lost (i.e., used for heating the water) in the first meter as a
result of the rapid absorption of the infrared part of the spectrum.

The surface layer of the ocean undergoes a temperature cycle
similar to that in the atmosphere. The surface water is warmer in
summer and colder in winter. Like the atmosphere, the seasonal temperature range is smaller in the tropics than in higher latitudes. There
are exceptions to the general rule, related to upwelling, seasonal
shifts of currents, and other local effects, but for much of the ocean
the observed seasonal temperature cycle can be approximated by
ignoring the transport term (Q_V) and considering only the seasonal
variations in the other four terms: incoming radiation from the sun
balanced against the heat losses of effective back radiation, evaporation, and sensible heat transfer. Thus the amount of heat available for
heating the ocean over any period of time t is

$$\int_0^t Q_T \, dt = \int_0^t (Q_s - Q_b - Q_e - Q_h) \, dt \qquad (3.4)$$

The distribution of that heat in the surface layer Z is

$$\int_0^t Q_T \, dt = \int_0^z c_P \rho \, \Delta T \, dz \tag{3.5}$$

where c_P is the specific heat of seawater, ρ the density, and ΔT the change in temperature. Although there have been a number of attempts to develop analytical expression for the right side of Eq. (3.5), they have been of limited success and usefulness. We shall content ourselves with simple descriptions. As a first example, let us assume that the sum of the heat losses does not change appreciably with season or time of day, and that the only significant seasonal and daily variation is in the incoming radiation (Figure 3.1). The total heat from the sun and the total heat loss from the ocean are the areas under the respective curves. There is a net gain of heat to the surface layer during the months of mid-February to mid-August and a net loss during the other six months (Figure 3.1a). From mid-February to mid-August the area under the incoming heat curve will be larger than that under the outgoing heat curve. During the remaining 6 months the situation will be reversed. Furthermore, the incoming radiation is cyclic during a 24-h period while the outgoing heat is approximately steady (Figure 3.1b).

The incoming radiation is absorbed throughout several tens of meters. All the processes responsible for heat loss take place at the surface. Thus, on the average, and perhaps nearly all the time over the entire ocean, there is a net loss of heat to the atmosphere in the top millimeter of the ocean (Figure 3.1c).

From Figure 3.1, a picture of the local variation in the temperature structure emerges. The minimum surface temperature in the ocean occurs in mid-February. As the incoming heat exceeds the outgoing heat, the ocean begins to warm. The maximum heat content of the ocean occurs in mid-August, at which point the daily heat losses exceed the heat gains.

Figure 3.2 is drawn from actual average temperature charts for the North Pacific. A seasonal thermocline develops during the summer. The depth of the thermocline can be explained at least approximately by considering the depth of penetration of the incoming radiation and the methods by which the heat can be mixed downward. Most of the mixing energy comes from the wind. The

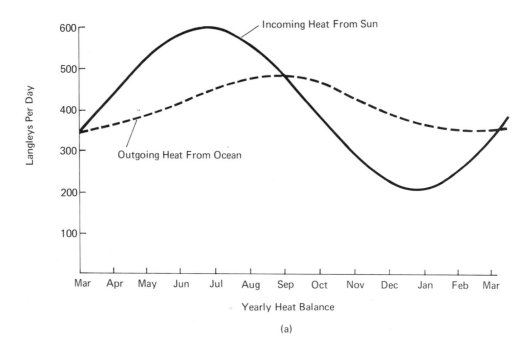

Yearly Heat Balance

(a)

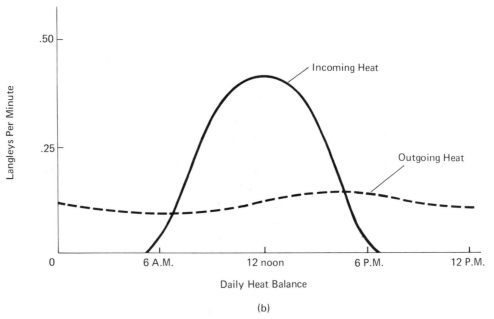

Daily Heat Balance

(b)

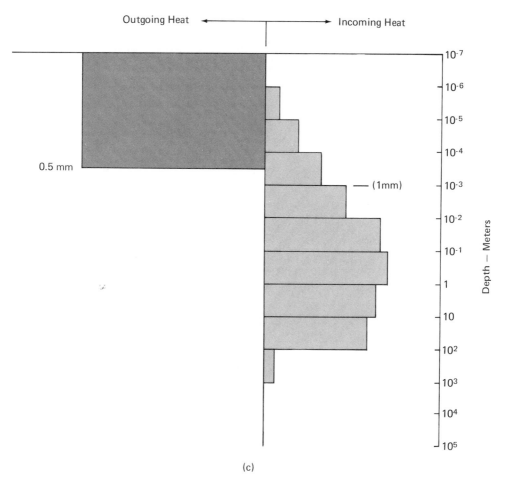

Figure 3.1. Schematic sketches of (a) seasonal variation in the incoming heat energy from the sun and the heat loss from the ocean at a typical mid-latitude site in the Northern Hemisphere; (b) diurnal variation in local heat balance; there is a heat loss over 24 h, which is balanced by a heat gain during daylight; (c) heat "sources" and "sinks" in the ocean surface layer. Note that the heat flux to the atmosphere is across the ocean–atmosphere viscous boundary layer while the heat from the sun is absorbed in the upper 100 m. As a result, there is always a net upward flux of heat in the top millimeter of the ocean.

thermocline is deeper in the spring than summer because the average winds are stronger in spring than summer and because the ocean becomes more stable as the thermocline becomes stronger. As stability increases, more energy is required to mix the heat deeper. Thus, as summer progresses, the thermocline becomes

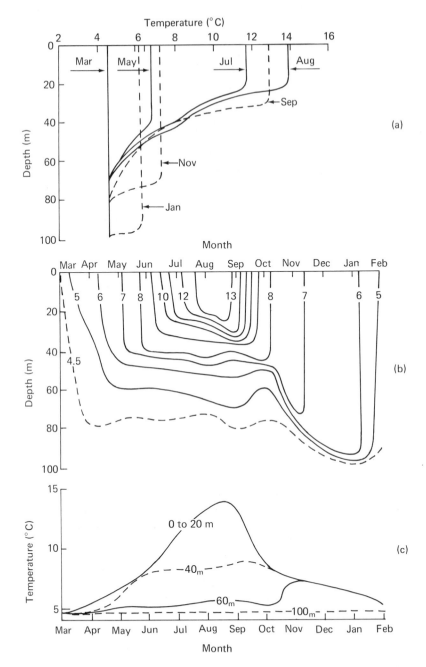

Figure 3.2. Typical growth and decay of the seasonal thermocline at a mid-latitude site in the Northern Hemisphere. The thermocline increases in strength and shallows as summer progresses and becomes deeper as it weakens in the fall. There is no seasonal change in the 100-m isotherm (data from 50°N, 145°W in the eastern North Pacific).

both shallower and stronger. In the fall, the thermocline weakens as the daily heat loss exceeds the heat gain. The combination of lower stability coupled with higher winds and increased convection (generated by sinking in the surface layer as the surface water cools) drives the thermocline deeper. By February the seasonal thermocline has disappeared and the process begins again.

It has not been possible to detect seasonal changes below 200 m. Whatever slight seasonal effect there might be at this depth is lost in the general noise level of small temperature fluctuations resulting from other oceanic processes. In much of the ocean the seasonal effect is gone by 100 m.

During the spring when there is a slight excess of incoming heat over heat loss, it is often possible to observe diurnal thermoclines (Figure 3.3). As can be seen in Figure 3.1b, there is a net heat gain during the day and a net heat loss at night. In some manner the excess heat remaining from these shallow diurnal thermoclines is mixed downward to form the seasonal thermocline.

Finally, it should be noted that as a consequence of the net heat loss at the surface there is a continual flux of heat from below to the surface. As a result, there is a negative temperature gradient in the surface skin of the ocean. On a calm night the temperature of the surface of the ocean that is radiating heat into space can be several tenths of a degree colder than the water a centimeter below the surface.

Heat Balance and Sea Ice

About 3–4% of the ocean is covered with ice. During winter the amount of incoming radiation from the sun is small, and during some periods of the long polar winter it is essentially zero. Because of the high reflectivity of ice, the percent of the incoming sun's radiation that is absorbed is considerably less than over water. The reflectivity (or albedo) of sea ice is between 30 and 40%, and the albedo can increase to as high as 95% directly after a freshly fallen snow. Incoming radiation that is not reflected is absorbed immediately in the top few millimeters of ice. Thus the only way in which heat can be transferred to the ocean is by conduction from the ice to the water; but since sea ice is always colder than the underlying water, the conduction of heat is *from* the water *to* the ice. Therefore, in regions of permanent sea ice the flux of heat is

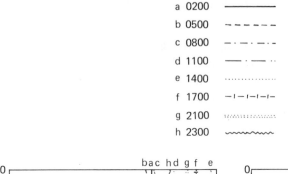

a	0200	————
b	0500	– – – – –
c	0800	–·—·—··
d	1100	——·——··
e	1400	·············
f	1700	–ı–ı–ı–ı–
g	2100	:::::::::::::::
h	2300	∿∿∿∿∿

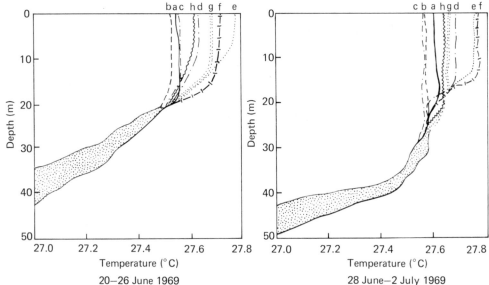

Figure 3.3. Growth and decay of a diurnal thermocline in spring. In a given 24-h period nearly all the heat that is gained during the day is lost at night. However, there is generally a small net heat gain each day. As a result, the average heat content of the second period is greater that that of the first period. (After Delnore, V.E., 1972: "Diurnal Variations of Temperature and Energy Budget for the Oceanic Mixed Layer During BOMEX," J. Phy. Oceanog., 2.)

always from the ocean to the ice and eventually to the atmosphere. Heat absorbed during the summer melts the surface ice and thus thins the ice layer.

The thickening and reducing of the ice layer is analogous to the formation and dissipation of the thermocline. In winter when the heat loss from the surface is great, the sheet thickens; in summer, the excess of incoming radiation is used to melt the ice. Near the fringes of the ice sheet the ice melts completely. In more polar regions, its thickness is reduced by melting and transpiration.

The rate at which an ice sheet thickens in winter in the absence of solar radiation is primarily a function of the temperature of the overlying atmosphere and the thickness of the ice. The growth of the ice is simply the heat loss to the atmosphere divided by the heat required to form a gram of ice:

$$\frac{dZ}{dt} = \frac{Q_h}{c_f \rho} \tag{3.6}$$

The heat flux to the atmosphere is determined by the thickness of the ice sheet and the difference in temperature between the atmosphere at the surface and the water at the bottom of the ice sheet where the freezing occurs:

$$Q_h = k_i \frac{\Delta T}{Z} \tag{3.7}$$

where k_i is the coefficient of thermal conductivity of the ice. Thus the rate of growth is inversely proportional to the thickness:

$$\frac{dZ}{dt} = \frac{k_i}{\rho c_f} \frac{\Delta T}{Z} \tag{3.8}$$

Ice forecasters use the equivalent of degree days in estimating the rate of growth of sea ice, where degree days are the product of the number of degrees below freezing and time. The growth rate of thickening can be found in terms of degree days (χ) by integrating Eq. (3.8) and assuming the underlying water is at zero degrees.

$$Z = \left(\frac{2k_i}{\rho c_f}\right)^{1/2} \chi^{1/2} \tag{3.9}$$

where $\chi = \int_0^t \Delta T \, dt$. Using values of $k_i = 5.5 \times 10^{-3}$ cal/degree-cm-s, $\rho = 0.92$ g/cm^3, and $c_f = 80$ cal/g

$$Z = 3.6\chi^{1/2} \text{ cm} \tag{3.10}$$

On the basis of Eq. (3.10) the maximum ice growth one would expect in a winter is about 3 m (Figure 3.4).

During the summer when the ice melts, Budyko reports that "the

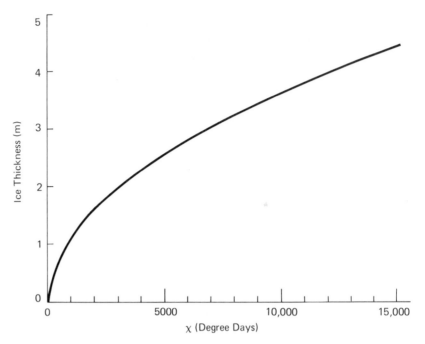

Figure 3.4. Growth of ice thickness as a function of degree days from Eq. (3.10).

total amount of melting per year in the Central Arctic amounts to several tens of centimeters.'' By comparison, he calculates the loss of about 2000 km³ of ice each year that drifts equatorward, a figure which would be equivalent to decreasing the ice thickness of the Central Arctic ice sheet by about 20 cm.

There are always open leads in the Arctic resulting from the differential movement of the ocean surface currents. Estimates of the amount of open water vary from 1–12%, with the strongest evidence toward the lower values. Assuming that the lower figure is correct, these open areas do not play a significant role in the heat balance of the Arctic basin.

Role of Advection in the Global Heat Balance

Averaged over a period of several years, the temperature of a given location in the atmosphere or ocean is nearly constant. Long-term, climatic warming or cooling trends are negligible compared to daily and seasonal changes. Since the heat balance for the earth as a whole is a radiation balance, the earth must radiate as much heat to space as it receives from the sun. An examination of that radiation balance (Figure 3.5) shows that, although there is a net balance integrated over the entire earth, there is a net gain of heat equatorward below approximately 40° and a net loss above 40°. Over the oceans, at least, the isolines are quite regular. Figure 3.5 indicates that there must be a net advection of heat by the ocean and the atmosphere from the tropics toward the poles with the common point at about 40°.

Under the assumption of no long-term change in temperature, Eq. (3.3) can be written

$$\int Q_V \, dt = \int (Q_s - Q_b - Q_e - Q_h) \, dt \tag{3.11}$$

where the integral is understood to be over at least a year. Figure 3.6 is a pictorial representation of the right side of Eq. (3.11) for the oceans alone. Positive terms imply an advection of heat from the area and negative values advection of heat into the area. The largest negative values are in the vicinity of the Gulf Stream and the Kuroshio, where warm water can be expected to be advected into the area. Figure 3.6 must be considered highly approximate. As Budyko noted, it is drawn by summing the individual terms, each of which has its own built-in error. Summing the difference between large numbers produces values with even larger uncertainties. Figure 3.6 strongly suggests, however, that advection plays a major role in the overall heat balance of the ocean. More critically, changes in the oceanic circulation, such as meanders in the Gulf Stream or a small shift in the strength or position of a major current, will effect the overall ocean–atmosphere heat balance. Thus changes in ocean circulation will result in changes in our weather patterns, and since the atmospheric winds drive the ocean currents, the combined ocean-atmosphere system is linked.

Global Heat Balance

Although there are daily and seasonal variations in the heat balance that result in the warming and cooling of the surface layer of the ocean, there must be a global heat balance on the average; otherwise, there would be a change in the average temperature of

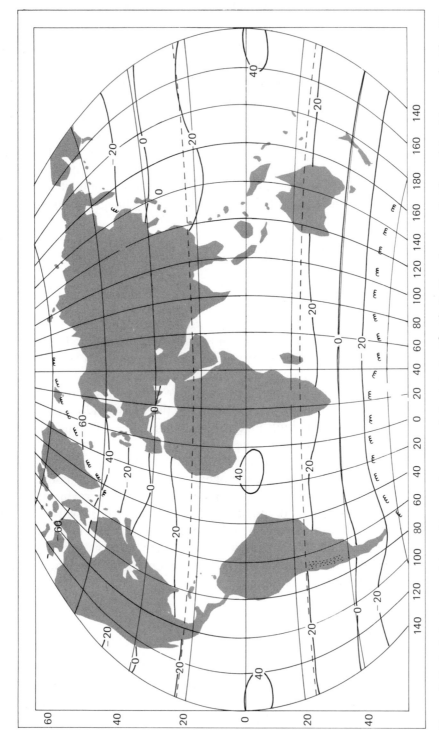

Figure 3.5. Radiation balance of the earth's atmosphere system (values in kcal cm^{-2} yr^{-1}). Positive values mean that the earth is gaining more heat than it is losing; negative values, the reverse. This figure implies a net flux of heat by the ocean and atmosphere from the tropics to the polar latitudes. The crossover point is about 40°N and S. (After Budyko, M. I., 1974: *Climate and Life*, Academic Press, New York.)

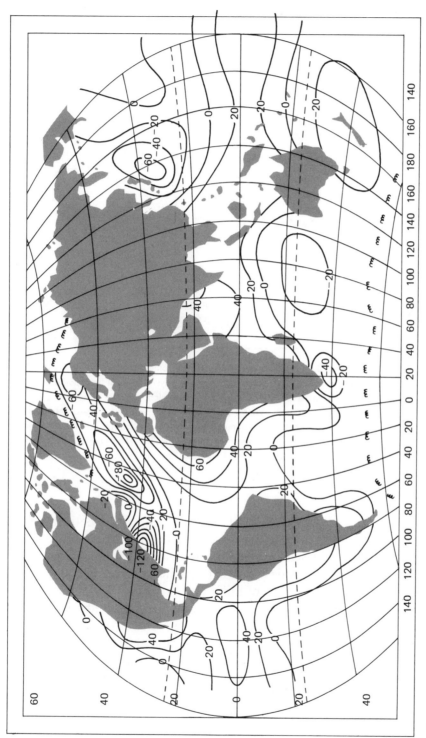

Figure 3.6 Heat flux across the ocean–atmosphere surface. Positive values mean a gain of heat to the surface water; negative values, the reverse (values in kcal cm^{-2} yr^{-1}). This figure implies a net advection of heat by ocean currents from positive-value areas to negative-value areas (see text for caveats). (After Budyko, M. I., 1974: *Climate and Life*, Academic Press, New York.)

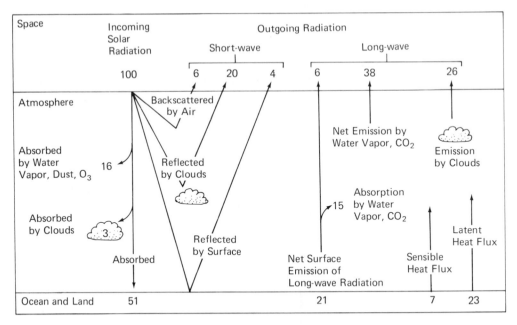

Figure 3.7. Partition of 100 units of incoming solar radiation showing percentages absorbed and scattered in various ways. The total heat balance requires 70 units of long-wave radiation from the earth along with the 30 units of backscattered shortwave radiation. Values are "whole earth, yearly averages." Note that of the 64 units absorbed by the atmosphere only 19 units (about 30%) come directly from the sun. (After U.S. Committee for the Global Atmospheric Research Program, 1975: *Understanding Climatic Change, a Program for Action,* National Academy of Science, Washington, D.C.)

the ocean and atmosphere. Figure 3.7 is a schematic attempt to show the total heat balance of the earth. The values are averaged over latitude and season, as well as land and ocean. Consideration of these averaged values, however, is instructive, because it shows the relative importance of different processes in maintaining the heat balance of the earth. The numbers are based upon 100 units of incoming radiation; the 100 units are equivalent to 0.5 ly/min.

Of the 100 units of radiation entering the atmosphere, about 3 units are absorbed by clouds and about 16 are absorbed by water vapor, aerosols, and the air molecules themselves. Some 30 units are backscattered and reflected into space. Of the original 100 units, slightly over 50% is available for heating the land, ocean, and ice. Of the 51 heat units from the earth's surface, 21 are long-wave radiation, 7 are sensible heat loss from the earth to the atmosphere, and 23 are a latent heat flux by evaporation.

From an examination of Figure 3.7, one learns that the albedo of the earth is about 30% (some 30 units of shortwave radiation are reflected back into space), and that the primary energy source for heating the atmosphere is the earth and not the sun. Of the incoming radiation, only 19 units are absorbed directly, compared with 45 units transferred from the earth to the atmosphere. One often speaks of the atmosphere as a gigantic heat engine driven by the sun, but what is often overlooked is that, although the sun is the primary energy source, 70% of this energy is transferred to the atmosphere from the earth. A comparatively small amount of the sun's energy is absorbed directly by the atmosphere.

It should be further noted that Figure 3.7 is calculated for the earth as a whole. Considering the ocean alone, it is estimated that two thirds of the heat transfer to the atmosphere is by evaporation, with long-wave radiation and sensible heat transfer reduced accordingly.

The details of the energy transfer between the earth and the atmosphere and how this balance is maintained contain the key to understanding our climate and much of our weather. The atmosphere derives its heat from the sun, but only about 30% of atmospheric energy is derived directly by absorbing energy from the sun. The rest comes in the form of effective back radiation, evaporation, and sensible heat transfer, and this heat transfer to the atmosphere occurs at night as well as day. To a much greater extent than the land surface, the ocean stores heat during the summer and releases it during the winter, and the ocean's ameliorating effect on the climate can be seen by comparing the seasonal and daily temperature ranges at coastal and inland stations. Inland stations at the same latitude show a much greater seasonal range in temperature than do coastal or island stations.

It is well known that there are year-to-year differences in the atmospheric climate; some winters are colder than others and some springs are wetter. The evidence for year-to-year changes in the temperatures and currents in the ocean is of more recent origin and more fragmentary. However, as suggested in Figure 3.6, advection apparently plays a major role in maintaining the global heat balance. It seems likely that the changes in year-to-year weather patterns are related to changes in the ocean, even if the causal relations connecting the two are still mostly conjectural. Figure 3.8 shows the anomaly of sea surface temperature in the North Pacific in the fall and winter of 1971. Arguments have been advanced that suggest that these anomalistic temperatures affected the prevailing weather patterns of the United States and Canada.

Of particular interest is the extent to which changing ocean conditions facilitate the formation of tropical cyclones (hurricanes in the Atlantic, typhoons in the Indian and Pacific oceans). It has been known for some time that the ocean is the major energy source of the hurricane. A hurricane loses its force as it moves over land or even over cold water. It has also been known for some time that there are certain tropical regions where hurricanes are most likely to form and that more form some years than others. What is less well understood is the exact combination of atmospheric and oceanic conditions that facilitate hurricane formation and growth.

Salt and Water Balance

In addition to an ocean heat balance, there must be a water balance and a salt balance. An evaporation rate of 126 cm/year is equivalent to removing 450×10^{12} m³/year or 0.03% of the total ocean volume each year. An equal amount of water reaches the ocean by rainfall and river runoff. It is estimated that 10% comes by way of rivers and the remainder by rainfall. Of the total amount of water on this planet, about 98% is in the ocean, 2% in glaciers, and only about 0.02% in freshwater lakes, rivers, and ground water. The total amount at any given time in clouds or water vapor is less than 1 part in 100,000. Table 3.2 estimates water balance by ocean as well as on a worldwide basis. The units are in centimeters per year. To convert into volume per year, multiply by the surface area of the ocean. Note that there is a net flow of water from the Pacific to the Atlantic and Indian oceans.

Although there is a global balance, there is, of course, no local balance. When ocean water evaporates, it leaves the salt behind; thus the remaining surface water becomes saltier. Charts of the surface salinity of the ocean demonstrate the effect of the local imbalance. In central ocean regions where evaporation exceeds precipitation, the surface salinity is higher than average. It is less than average in those

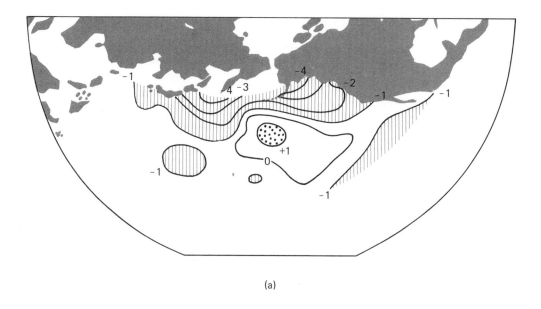

(a)

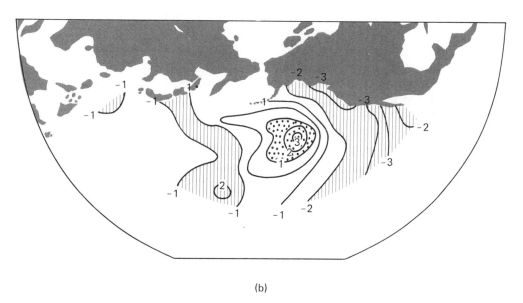

(b)

Figure 3.8. The difference in sea surface temperature in the North Pacific from a 20-year mean (1947–1967) for (a) the 1971 fall, and (b) 1971–1972 winter. Values are in °F. (After Namias, J., 1972: "Experiments in Objectively Predicting Some Atmospheric and Oceanic Variables for the Winter of 1971–1972," J. *Applied Meteo.*, 11.)

Table 3.2 Precipitation and evaporation for the world's oceans and river runoff in centimeters per year[*]

Ocean[†]	Precipitation (cm yr^{-1})	Evaporation (cm yr^{-1})	Runoff (cm yr^{-1})
Atlantic	89	124	23
Pacific	133	132	7
Indian	117	132	8
World ocean	114	126	12
Global transport (km^3/yr)	41×10^4	45×10^4	4×10^4

[*]After Budyko, M. I., 1974: *Climate and Life*, Academic Press, New York.
[†]To find individual oceanic values in terms of transport, multiply by surface area of ocean.

areas where the reverse is true. In general, coastal regions have lower salinities than the open ocean because of the influence of river runoff. It has been possible to show at least an approximate quantitative relationship between observed surface salinity and the estimated differences between evaporation and precipitation. Presumably, the fact that the Atlantic is more saline than the Pacific can be explained at least in part by the oceanic balance of Table 3.2.

That one can approximately predict the surface salinity by only considering the exchange of freshwater suggests that the salt balance problem is of smaller magnitude. It is estimated that the rivers bring in about 3×10^{12} kg of dissolved solids per year. Thus, although the global water balance requires consideration of only about 0.03% of the total ocean volume, the ocean salt balance each year involves less than 10^{-7} of the total salt content of the ocean. Even if all the salt coming into the ocean were to be added to seawater, which we know is not the case, it would be about 500 years before we could detect an increase in the average salinity with our present measurement techniques.

chapter 4

CONSERVATION EQUATIONS, DIFFUSION, AND BOX MODELS

To examine further the oceanic distribution of salt, heat, and other properties, it is necessary to consider the conservation equation in a more formalistic manner. In the process we shall also consider the notion that material can be transported by diffusive processes as well as by ocean currents.

Conservation Equation

Consider an imaginary small rectangular volume or box (V) whose six sides are $B_1, B_2 \ldots, B_6$, and that this volume is imbedded in a fluid of varying density and velocity (Fig. 4.1). It can be shown that the rate of change in density $\delta\rho/\delta t$ within the box is related to the sum of the flux of material across the six sides in the following way:

$$V \frac{\delta\rho}{\delta t} = -\sum_{i=1}^{6} B_i \rho_i V_i \qquad (4.1)$$

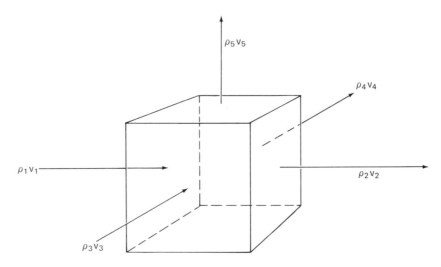

Figure 4.1. In the absence of any *sources* or *sinks* within the box, the sum of the values of $(\rho_i V_i)$ flowing into the box must equal that flowing out.

where ρ_i and V_i are, respectively, the density and the velocity component normal to the side of B_i of the box, and if we adopt the convention that flow into the volume is negative and flow out is positive.

A similar summation equation can be written for the change of salinity*:

$$V\,\frac{\delta S}{\delta t} = -\sum_{i=1}^{6} B_i S_i V_i \tag{4.2}$$

If the flux of mass (or salt) into the box equals that flowing out, the left side of the equation is zero, and one can simply write

$$\Sigma\, B_i \rho_i V_i = 0 \tag{4.3}$$

$$\Sigma\, B_i S_i V_i = 0$$

*Since salinity is measured in grams per kilogram rather than grams per liter, Eq. (4.2) should be written $V(\delta S \rho/\delta t) = -\Sigma\, B_i S_i \rho_i V_i$. The distinction is seldom made in these kinds of calculations; rather, a new definition of salinity in units of grams per liters is assumed. The difference is not important for any numerical calculations.

Explicit in these formulations is the idea that there can be no change in salinity or density in the volume except by flux of material in or out. Such properties are called *conservative*, and similar equations can be written for other conservative properties. *Nonconservative* properties are those whose value can change independently of the flux. Examples are radioactive materials such as tritium or biologically active materials such as dissolved oxygen or phosphate. Typical conservation equations for nonconservative properties are

$$V \frac{\delta H}{\delta t} = -\Sigma \ B_i H_i V_i - V \gamma H \tag{4.4}$$

$$V \frac{\delta O}{\delta t} = -\Sigma \ B_i O_i V_i - V \epsilon \tag{4.5}$$

where γ is the half-life of the radioactive material, 12.5 years in the case of tritium, and ϵ denotes the rate at which dissolved oxygen is being locally produced by photosynthetic processes or depleted by oxidation processes. A conservative property differs from a nonconservative property in that the former has no *sources* or *sinks* within the medium. Although dissolved oxygen may be "quasi conservative," below the surface layer it would appear that oxidation processes in the deep ocean account for a decrease per year of about 0.0015 milliliter (ml) of oxygen per liter of seawater.

Conservation Equation: Differential Form: Although writing the conservation equations in terms of flow in and out of an imaginary box is satisfactory for many purposes, these equations are most often written in their differential form, and it is thus that they are derived (see Appendix I). In terms of a rectilinear coordinate system with axes x, y, and z, and corresponding velocity components u, v, and w, the conservation of mass can be written

$$\frac{\partial \rho}{\partial t} = -\frac{\partial}{\partial x} (\rho u) - \frac{\partial}{\partial y} (\rho v) - \frac{\partial}{\partial z} (\rho w) \tag{4.6}$$

and the conservation of salt is given approximately by

$$\frac{\partial S}{\partial t} = -\frac{\partial}{\partial x} (Su) - \frac{\partial}{\partial y} (Sv) - \frac{\partial}{\partial z} (Sw) \tag{4.7}$$

Similar equations can be written for other conservative properties. The analogous equations for nonconservative properties, are

$$\frac{\partial H}{\partial t} = -\frac{\partial}{\partial x}(Hu) - \frac{\partial}{\partial y}(Hv) - \frac{\partial}{\partial z}(Hw) - \gamma H \tag{4.8}$$

$$\frac{\partial O}{\partial t} = -\frac{\partial}{\partial x}(Ou) - \frac{\partial}{\partial y}(Ov) - \frac{\partial}{\partial z}(Ow) - \epsilon \tag{4.9}$$

For purposes of these conservation equations, seawater can be assumed incompressible, which simplifies Eqs. (4.6) through (4.9):

$$\frac{\partial u}{\partial x} + \frac{\partial v}{\partial y} + \frac{\partial w}{\partial z} = 0 \tag{4.10}$$

$$\frac{\partial S}{\partial t} = -u\frac{\partial S}{\partial x} - v\frac{\partial S}{\partial y} - w\frac{\partial S}{\partial z} \tag{4.11}$$

$$\frac{\partial O}{\partial t} = -u\frac{\partial O}{\partial x} - v\frac{\partial O}{\partial y} - w\frac{\partial O}{\partial z} - \epsilon \tag{4.12}$$

For many purposes it can be assumed that the ocean is in a "steady state"; that is, there is no appreciable change in the month-to-month or year-to-year observed values of such properties as salinity, temperature, and oxygen. For steady-state conditions, the left sides of Eqs. (4.11) and (4.12) are zero.

Equations (4.6) through (4.12) are *partial differential equations* Although no attempt is made in this text (except in Appendix I) to manipulate such equations, it is important for the reader to at least understand the basis of notation. When a property has only one dependent variable, the usual differential form is dw/dt, which means that the independent variable w is a function of t and of t alone. When the independent variable is a function of more than one dependent variable, it is necessary to be explicit in indicating how the differentiation is done. For example, the value of salinity in an area may vary with depth z as well as in the horizontal plane (x and y), and it may vary with time t. Thus salinity is a function of x, y, z, and t; and if one wishes to know the "salinity gradient," one must be explicit in indicating whether he (or she) wants to know the gradient in terms of change of salinity with depth or in the x or y direction, or in some intermediate direction. Similarly, if one wishes to know how salinity changes in time, one must also specify whether or not he (or she) means the change in time at a given point in space.

The partial differential $\partial S/\partial x$ is more properly written $(\partial S/\partial x)_{y,z,t}$, which means the change of salinity in the x direction at a specified and constant value of y, z, and t. After one is introduced to partial differential notation, the subscripts are usually omitted; but in Eq. (4.6), for example, the first term means the change of density with time at a given point (i.e., holding x, y, and z constant); the second term means the change of ρu in the x direction, holding y, z, and time constant; and so on.

Diffusion

Until now we have tacitly assumed that changes in the distribution of properties occur only by advection. Yet elsewhere reference has been made to mixing processes, which at least implies that material can be transported by ways other than advection. Imagine the following experiments. A salt plug is dissolved on the bottom of a large tank of freshwater. Even though there are no currents in the tank, the salt will gradually diffuse through the tank, and the salinity at a point some distance from the dissolving plug will gradually increase. After a very long time, the salt will be completely diffused through the tank and the salinity uniform.

As a second example, imagine a layer of warm water resting on a layer of cold water with a very sharp temperature gradient between the two. In the absence of other currents, the temperature gradient across the interface will weaken as heat is diffused from the upper to the lower layer.

Diffusion of heat, salt, and momentum can be by molecular processes. The molecules within the fluid are in continuous motion and interacting with one another. Thus, even though the fluid is motionless, transfer can take place at the molecular level.

The flux of heat, salt, and momentum is proportional to the gradient of the property:

diffusion of salt $\qquad\qquad F = -j\,\dfrac{ds}{dn}$ gm/cm^2 s

diffusion of heat $\qquad\qquad Q = -\,kc_p\dfrac{dT}{dn}$ cal/cm^2 s $\qquad$ (4.13)

diffusion of momentum $\qquad \tau = -\mu\,\dfrac{dv}{dn}$ gcm s^{-1}/cm^2 s

where n is normal to lines of constant temperature, salinity, and velocity. The negative signs mean the diffusion occurs down the gradient. For water, the coefficient of

molecular diffusivity $\qquad (j) \cong 2 \times 10^{-5}$ g/cm s

molecular conductivity $\qquad (k) \cong 1 \times 10^{-3}$ g/cm s

molecular viscosity $\qquad (\mu) \cong 2 \times 10^{-2}$ g/cm s

and the specific heat at constant pressure $c_p \cong 0.94$ cal/g °C. Given these constants and the known average gradients in the ocean, it is easy to calculate that molecular diffusion of heat, salt, and momentum is very small compared to the diffusion that would appear to be occurring in the ocean. For example, the heat loss by conduction from the ocean to the atmosphere is estimated at about 25 ly/day (Chapter 2), or about 3×10^{-4} cal/cm^2 s. According to Eq. (4.13), this translates into the temperature increasing with depth at a rate of about 0.3°C/cm. Such gradients may exist in the top few millimeters of the ocean where molecular diffusion is important, but below the interface a temperature gradient as large as 3×10^{-3} °C/cm is rare.

In the ocean, measurable diffusion occurs by turbulent processes. It is argued that the eddies and swirls that occur in turbulent motion are analogous to molecules, and their movement is analogous to the mean free path of molecules. By arguing that eddy diffusion, eddy conductivity, and eddy viscosity are physically analogous to these processes at the molecular level, one can write equations identical with Eq. (4.13) by replacing the molecular coefficients with eddy coefficients.

Turbulent eddies in the ocean have scales that vary from tens of millimeters to tens of kilometers. As a consequence, a wide range of eddy coefficients can be expected. Values range in the order of

$A_z = 1–10$ g/cm s

$A_h = 10^3–10^7$ g/cm s

where the subscripts h and z denote horizontal and vertical. Because the ocean is stably stratified, vertical turbulence is greatly depressed compared to horizontal turbulence.

Also note that a vertical turbulent coefficient of 10 g/cm s is 10^4 times the value of the molecular conductivity coefficients. According to Eq. (4.13), this means that the vertical temperature gradient necessary to transfer heat by turbulent diffusion need only be $\frac{1}{10000}$ the gradient required to transfer the same amount of heat by molecular conductivity.

Allowing for the possibility of the flux of material by turbulent diffusion processes requires the addition of a set of terms to the conservation equations for both conservative and nonconservative properties. In our "box model" form, they are

$$V \frac{\delta S}{\delta t} = -\Sigma \, B_i S_i V_i + \Sigma \, A_i B_i \left(\frac{\partial S}{\partial n} \right)_i \tag{4.14}$$

$$V \frac{\delta O}{\delta t} = -\Sigma \, B_i O_i V_i + \Sigma \, A_i B_i \left(\frac{\partial O}{\partial n} \right)_i - \epsilon V \tag{4.15}$$

where $(\partial O/\partial n)_i$ and $(\partial S/\partial n)_i$ means the gradient of oxygen and salinity normal to the plane B_i.

In differential form, these equations are generally written (see Appendix I)

$$\frac{\partial S}{\partial t} = -\frac{\partial}{\partial x} (Su) - \frac{\partial}{\partial y} (Sv) - \frac{\partial}{\partial z} (Sw)$$
$$+ \frac{A_h}{\rho} \left(\frac{\partial^2 S}{\partial x^2} + \frac{\partial^2 S}{\partial y^2} \right) + \frac{A_z}{\rho} \left(\frac{\partial^2 S}{\partial z^2} \right) \tag{4.16}$$

$$\frac{\partial O}{\partial t} = -\frac{\partial}{\partial x} (Ou) - \frac{\partial}{\partial y} (Ov) - \frac{\partial}{\partial z} (Ow)$$
$$+ \frac{A_h}{\rho} \left(\frac{\partial^2 O}{\partial x^2} - \frac{\partial^2 O}{\partial y^2} \right) + \frac{A_z}{\rho} \left(\frac{\partial^2 O}{\partial z^2} \right) - \epsilon \tag{4.17}$$

Box Models and Mixing Times

There are a series of ocean problems that can be examined by dividing the oceans into a series of separate basins or layers (boxes) and considering the flux of material into and out of these boxes. Consider a simple box of volume V filled with water of constant density and salinity. Water is "flowing" into and out of the box by advection and/or diffusion. Under steady-state conditions we can write for the conservation of mass and salt

$$-\Sigma\, B_i \rho_i V_i + \Sigma\, A_i B_i \left(\frac{\partial \rho}{\partial n}\right)_i = 0 \tag{4.18}$$

$$-\Sigma\, B_i S_i V_i + \Sigma\, A_i B_i \left(\frac{\partial S}{\partial n}\right)_i = 0 \tag{4.19}$$

The units of $B\rho V$ are mass divided by time, which is the *mass transport*. The units of BV are volume per unit time, which is called the *volume transport*. Because the density of seawater is nearly 1, the numerical value of mass transport in and out of the box in tons per second is nearly identical with that of volume transport in cubic meters per second. The symbol T is often used to designate either in oceanography. Furthermore, the term volume transport or mass transport is often used to subsume any flux of material caused by eddy diffusion processes. As a result, Eqs. (4.18) and (4.19) can be simplified to

$$\Sigma\, T_i = 0 \tag{4.20}$$

$$\Sigma\, T_i S_i = 0 \tag{4.21}$$

It is often useful to consider the conservation of two properties together. For example, consider the Mediterranean as an isolated basin in which the outflowing water has a uniform salinity of 37.75‰ connected to an ocean of salinity 36.25‰ (Figure 4.2). It is known that water flows in and out of the basin, but not how much. It is, however, known (or at least estimated) that the excess of evaporation over precipitation is about 70×10^3 tons/s. It is possible to write two conservation equations, conservation of water (or mass) and conservation of salt. The first is simply

$$T_{\text{in}} = T_{\text{out}} + 70{,}000 \text{ tons/s} \tag{4.22}$$

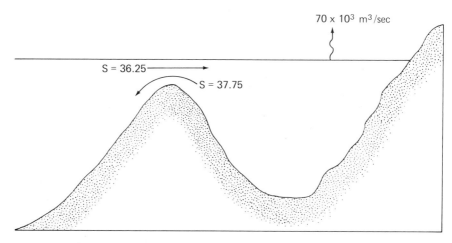

Figure 4.2. Net flow in and out of the Mediterranean across the Straits of Gibralter can be calculated knowing the salinity of the two flows and the flux of water across the ocean–atmosphere interface.

The second is closely approximated by assuming the product of salinity and transport

$$T_{in}S_{in} - T_{out}S_{out} = 0 \text{kg/s} \tag{4.23}$$

since salinity is approximately in units of kilograms of salt per cubic meter of water. With two equations and two unknowns it is possible to solve for T_{in} and T_{out}, which are 1.75 and 1.68×10^6 tons/s (m³/s), respectively. Calculations similar to these were first made by Schott in 1915. Others have attempted to solve more complicated models with more basins and more unknowns. Every additional unknown requires an additional conservation equation.

Finally, consider the question of how long a particle of water stays in a basin such as the Mediterranean once it gets there. Clearly, water is being continually exchanged across the Straits of Gibraltar and with the atmosphere by evaporation and precipitation. Some particles will be exchanged quickly, and others will remain in the Mediterranean for a very long time; but it is useful to have some way of characterizing the average time a molecule will spend in the basin. Under certain assumptions this can be done easily.

The Mediterranean basin problem and other similar ones can be

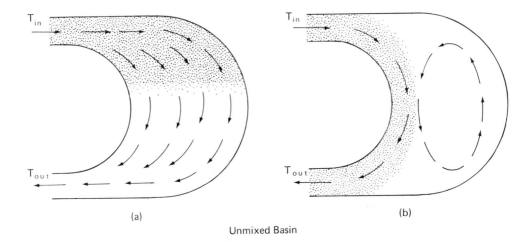

(a)

Unmixed Basin

(b)

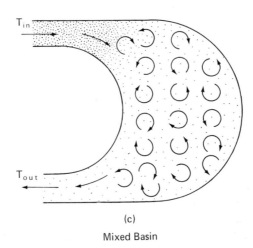

(c)

Mixed Basin

Figure 4.3. Various extremes of mixing in a flow-through system. In (a) there is no mixing and the new water flowing in simply replaces the water flowing out. There is no mixing in (b) either. In this case the water that flows in also flows out without mixing with the original water in the basin. In (c) the mixing is complete and "instantaneous."

simplified to a single flow into the basin and a single flow out (although in fact the single flow may be the sum of a number of local inputs).

Three rather extreme examples can be considered. One might assume that the total inflow (T) gradually, but systematically, replaces the volume of water V in the basin. The time (Ξ) for this to occur (Figure 4.3a) is

$$\Xi = \frac{V}{T} \tag{4.24}$$

A second possibility is that the same water flows in and out and that most of the water in the basin is undisturbed (Fig. 4.3b). In this case an infinite time is required to replace all but a small fraction of the water in the basin. The third extreme is that the incoming water mixes completely and immediately with all the water in the basin (Fig. 4.3c). In this case the residence time of some of the water particles is very short, and some very, very long.

The assumption of complete mixing does allow one to define a statistical residence time. It can be shown that the flushing time (or residence time) defined by Eq. (4.24) defines, in the case of complete mixing, the time necessary for all but $1/e$ or $1/2.73$ of the molecules in the original volume of water to be replaced (see Appendix I). The volume of the Mediterranean is 4×10^{15} m³. Assuming that the mixing of a particle of water in the Mediterranean is comparatively rapid, the average residence time of a particle of water in the Mediterranean is

$$\Xi = \frac{4 \times 10^{15} \text{ m}^3}{1.75 \times 10^6 \text{ m}^3/\text{s}} = 2.2 \times 10^9 \text{ s} \cong 70 \text{ years}$$

Calculations such as the above are straightforward. Understanding their significance, in light of Figure 4.3, is sometimes more difficult.

chapter 5

EQUATION OF MOTION

In principle, the relationships explaining motion in the ocean can be written in a series of mathematical equations. In practice, we can only evaluate these equations in part; thus those equations most commonly used in representing currents, waves, tides, turbulence, and other forms of motion are at best fair approximations of what occurs. In this chapter we consider the equations of motion and how they apply to understanding some aspects of ocean circulation.

Any quantitative discussion of forces and motions requires a coordinate system. The system most commonly used in oceanography is the rectilinear Cartesian system in which the earth is assumed to be flat. A spherical coordinate system would be more realistic, but it is also more complicated. The Cartesian system has been found adequate for nearly all problems in physical oceanography.

The usual convention is to assume a plane in which the x axis points east, the y axis points north, and the z axis is up; more precisely, the z axis is in the direction opposite to the gravitational vector. The corresponding velocity vectors are u, v, and w.

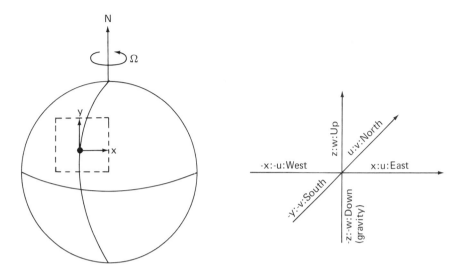

Figure 5.1. Cartesian coordinate system used in this text.

Although meteorologists and oceanographers may agree on the coordinate system in Figure 5.1, they use different conventions for describing winds and currents. A north current is current flowing toward the north; a north wind is a wind blowing from the north. The convention is confusing, but there is little likelihood that it will be changed. To minimize the confusion, this text refers to northerly winds and northward currents.

Newton's second law states that the acceleration of a particle is proportional to the sum of the forces acting on the particle.

$$\frac{du}{dt} = \frac{1}{m} \Sigma \text{ F} \tag{5.1}$$

In discussing fluid motion, the relationship is usually written

$$\frac{du}{dt} = \frac{1}{\rho} \Sigma \text{ F} \tag{5.2}$$

where it is now understood that the forces are "per unit volume," since Eq. (5.2) can be derived from Eq. (5.1):

$$\frac{du}{dt} = \frac{V}{m} \Sigma \frac{F}{V}, \qquad \rho = \frac{m}{V} \tag{5.3}$$

As written, Eqs. (5.1) and (5.2) apply to the components of the forces acting in the east-west or x direction. Similar equations can be written for the force components acting along the other two axes:

$$\frac{du}{dt} = \frac{1}{\rho} \Sigma F_x$$

$$\frac{dv}{dt} = \frac{1}{\rho} \Sigma F_y \tag{5.4}$$

$$\frac{dw}{dt} = \frac{1}{\rho} \Sigma F_z$$

There are four important kinds of forces acting on a fluid particle in the ocean: gravitational, pressure gradient, frictional, and Coriolis. In a generalized way, Eq. (5.2) may be written

$$\text{density} \times \text{particle acceleration} = \text{gravity} + \text{pressure gradient} + \text{Coriolis} + \text{friction} \tag{5.5}$$

The mathematical expression for the forces of gravity, pressure, and Coriolis may be expressed quite simply. The various forms of the frictional forces are less easy to express in a precise manner, and they are considerably more difficult to measure in the ocean. The two problems are not unrelated. Note that, by choosing a coordinate system such that the z axis points opposite to the direction of gravity, there will be no gravitational force in either the x or y direction.

Acceleration

Before looking at the various force terms, it is necessary to indicate something about the acceleration of fluids. Newton's second law is usually introduced in terms of particle mechanics (a block sliding down an inclined plane or the movement of billiard balls). As written,

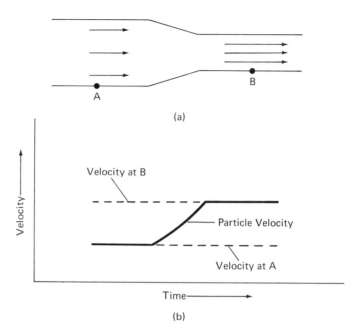

Figure 5.2. (a) As the channel narrows, the water must flow faster. A current meter at point *A* or *B* records a steady velocity. However, the velocity of a particle increases from *A* to *B*. (b) Local acceleration is zero; the average particle acceleration between *A* and *B* is not.

Eq. (5.4) applies to the motion of a particle. In continuum mechanics (or fluid mechanics) there are two kinds of acceleration for which there are ready operational definitions. The first is called *particle acceleration* and the second *local acceleration* Although one can be defined in terms of the other, the two are not identical. Consider the example in Figure 5.2. A constant volume of water is being forced to flow through a channel that becomes increasingly narrow. A fluid particle passing through the channel is being accelerated as it moves from *A* to *B*. However, a current meter inserted at any point along the channel *records a constant average velocity* Thus, in this example, the local acceleration is zero.

In many problems it is desirable to write Eq. (5.4) in terms of the local acceleration rather than the particle acceleration. The two are related in the following way (see Appendix I):

particle acceleration = local acceleration + field acceleration terms

$$\frac{du}{dt} = \frac{\partial u}{\partial t} + u\frac{\partial u}{\partial x} + v\frac{\partial u}{\partial y} + w\frac{\partial u}{\partial z}$$

$$\frac{dv}{dt} = \frac{\partial v}{\partial t} + u\frac{\partial v}{\partial x} + v\frac{\partial v}{\partial y} + w\frac{\partial v}{\partial z} \qquad (5.6)$$

$$\frac{dw}{dt} = \frac{\partial w}{\partial t} + u\frac{\partial w}{\partial x} + v\frac{\partial w}{\partial y} + w\frac{\partial w}{\partial z}$$

or, in general,

$$\frac{d}{dt} = \frac{\partial}{\partial t} + u\frac{\partial}{\partial x} + v\frac{\partial}{\partial y} + w\frac{\partial}{\partial z} \qquad (5.7)$$

Pressure Gradient

Of the various terms in Eq. (5.5), perhaps the pressure gradient is the easiest to visualize. A particle will move from high pressure to low pressure, and the acceleration is simply proportional to the pressure gradient. A mechanical analog is a ball on a frictionless inclined plane. The ball rolls down the plane (from high to low pressure), and the acceleration of the ball is proportional to the inclination of the plane (pressure gradient). Mathematically, Eq. (5.5) now becomes (see Appendix I)

$$\frac{du}{dt} = -\frac{1}{\rho}\frac{\partial p}{\partial x} + \text{other forces}$$

$$\frac{dv}{dt} = -\frac{1}{\rho}\frac{\partial p}{\partial y} + \text{other forces} \qquad (5.8)$$

$$\frac{dw}{dt} = -\frac{1}{\rho}\frac{\partial p}{\partial z} + \text{other forces}$$

Pressure gradients arise in a variety of ways. One of the simplest is by

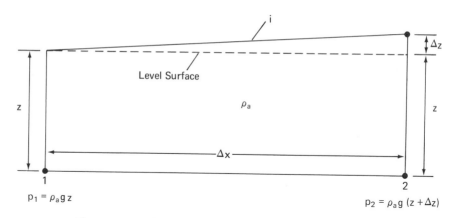

Figure 5.3. Slope of the sea surface creates a pressure gradient, Eq. (5.10).

a sloping water surface. Imagine a container with an ideal fluid whose density is ρ_a, and that in some manner it is possible to have the water surface slope as in Figure 5.3 without causing any other motion. Remembering that the pressure at any point in the fluid is simply the weight,

$$p_1 = \rho_a g Z$$

$$p_2 = \rho_a g (Z + \Delta Z)$$

$$(5.9)$$

The resulting pressure-gradient term is

$$\frac{1}{\rho}\frac{\partial p}{\partial x} = \frac{1}{\rho_a}\frac{p_2 - p_1}{\Delta X}$$

$$= g\,\frac{\Delta Z}{\Delta X}$$

$$(5.10)$$

$$= gi$$

where i is the slope of the fluid surface.

It can easily be shown that the horizontal pressure gradient is identical everywhere within the fluid. Thus, if there were no other forces acting, Eq. (5.8) says that the entire fluid would be uniformly accelerated toward the lower pressure.

Coriolis Force

The Coriolis "force" is the most difficult of the four forces to comprehend because physical intuition is of little avail. Most of us have some qualitative ideas of what to expect from the forces of gravity, pressure, and friction; but there is little in our experience to indicate what happens to a particle under the influence of the Coriolis force.

The first thing to understand about the Coriolis force is that it is not a true force at all; rather, it is a device for compensating for the fact that the particle which is being accelerated by the forces of gravity, pressure, and friction is being accelerated on a rotating earth, and all measurements of forces and accelerations are made relative to a rotating earth. Two examples will indicate the nature of the problem. The earth, with a radius of about 6400 km which rotates once every 24 h, has tangential velocities as indicated in Figure 5.4. Let us assume that a particle of water is set in motion at 45°N with a southward velocity of 1m/s, and that no forces other than gravity act upon it. According to Newton's first law, a particle in motion will continue to move at a constant velocity in the absence of any force. Thus, in slightly less than 2 days the particle should pass 30°N, continuing its southward journey at 1 m/s. However, the 1-m/s velocity is measured

Figure 5.4. Because of the change of tangential speed with latitude, a particle moving toward the equator appears to be accelerated to the west.

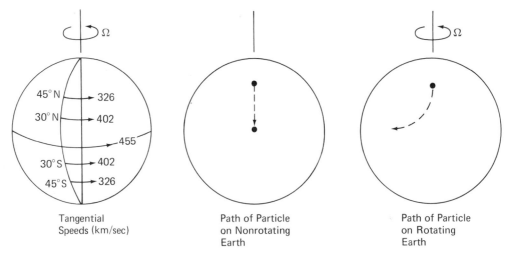

Tangential Speeds (km/sec)

Path of Particle on Nonrotating Earth

Path of Particle on Rotating Earth

relative to the earth. In terms of a coordinate system that allows for earth rotation, the particle also has an eastward velocity of 326 m/s. However, the eastward tangential velocity at 30°N is 402 m/s. Thus to an observer on the earth it would appear that the particle not only has the 1-m/s southward component, but it would also be traveling westward at a speed of 76 m/s. To an observer on the earth it would appear that the particle had undergone a tremendous westward acceleration.

One can play the same game by starting the particle northward at 30°N at 1 m/s and find that at 45°N the particle is now apparently traveling eastward at 76 m/s. You can do the same thing between 30° and 45°S and find that the east-west velocities are of the same magnitude, but that the apparent acceleration is in the opposite sense. As you move your imaginary particle north and south along the earth you can find a general rule: in the Northern Hemisphere the apparent acceleration is always to the right of the direction in which the particle is moving; in the Southern Hemisphere the apparent acceleration is to the left of the direction of flow. At the equator, the acceleration will go through an inflection point; there will be no acceleration at the equator.

For a second example, consider a pendulum suspended at the North Pole and free to swing in any direction. Assume that at 12 noon it is set in motion such that it is swinging along the 90°E–90°W longitude axis (Figure 5.5). In the absence of other forces, it will continue to swing in the same direction as the earth rotates under it. Looking down on the North Pole, the earth rotates counterclockwise

Figure 5.5. A pendulum at the north pole will swing in its original plane. To an observer on the earth, which is rotating beneath the pendulum, it appears that the pendulum is rotating clockwise at the rate of 15°/h.

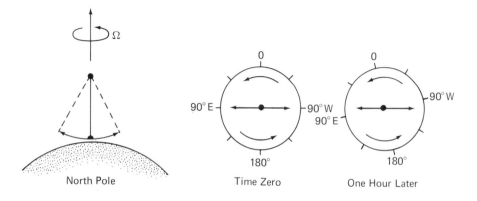

15°/h. To an observer on the earth standing near the pole, the pendulum appears to rotate *clockwise* 15°/h. In 12 h the pendulum will have rotated such that it is again swinging along the 90°E–90°W axis.

A similar result occurs at the South Pole, only for an observer in space looking down upon the South Pole, the earth appears to be rotating clockwise. Thus to an observer on the earth standing near the pole, the pendulum will appear to rotate *counterclockwise*

Imagine now a similar pendulum swinging along the east–west axis at the equator. As the earth rotates under it, the pendulum will continue to swing along the east–west axis. It can be shown that the time required for the pendulum to rotate 180° so that it is swinging in the original plane is

$$T = \frac{12\ \text{h}}{\sin\ \phi} \tag{5.11}$$

where ϕ is latitude. At 90° latitude the period is 12 h; at the equator the period is infinity. Such a pendulum is called a *Foucault pendulum,* after Jean Bernard Foucault, who demonstrated such a pendulum in Paris in 1851.

As the two examples demonstrate, there is a class of problems for which it is necessary to allow for the effect of the earth's rotation. The simplest way to allow for the earth's rotation is to add an apparent force, the Coriolis force. In our coordinate system this force is given with sufficient accuracy for most oceanographic problems by writing Eq. (5.5) as

$$\frac{du}{dt} = -\frac{1}{\rho}\frac{\partial p}{\partial x} + vf + \text{other forces} \tag{5.12}$$

$$\frac{dv}{dt} = -\frac{1}{\rho}\frac{\partial p}{\partial x} - uf + \text{other forces}$$

where $f \equiv 2\Omega \sin \phi$, and Ω is the angular velocity of the earth, $2\pi/24$ h or $7.29 \times 10^{-5}\ \text{s}^{-1}$.

A complete derivation of the Coriolis force is best done with vector algebra (see Appendix I). However, some insight can be gained by considering the centrifugal acceleration of a particle on the surface of the earth, which is

$$\frac{U^2}{q} \tag{5.13}$$

where $U = \Omega q$ (Figure 5.6). If the particle is moving in an eastward direction u, the centrifugal acceleration is

$$\frac{(U + u)^2}{q} = \frac{U^2}{q} + \frac{2Uu}{q} + \frac{u^2}{q} = \Omega^2 q + 2\Omega u + \frac{u^2}{q} \tag{5.14}$$

Since U is the order of a thousand times greater than u, the first term is the order of a thousand times larger than the second, which is of the order of a thousand times larger than the third. This last term is small enough to be ignored. The second term is the Coriolis acceleration.

As can be seen in Figure 5.6, the direction of the Coriolis acceleration can be resolved into two components, one normal to the plane of the earth and the other parallel to the plane of the earth. The latter has the value $2\Omega u \sin \phi$ and is the horizontal component of the Coriolis acceleration that applies to east–west motion. Note that, if the particle were moving westward, the same absolute value would apply, but the second term in Eq. (5.14) would be negative. The

Figure 5.6. Eastward velocity u relative to the surface of the earth increases the centrifugal acceleration of the particle relative to the surface of the earth (see text for argument).

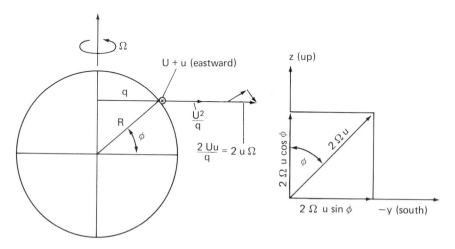

vector in Figure 5.6 would be pointed toward the earth's axis, and the component in the plane of the earth would be pointing north.

An analysis of Eq. (5.12) indicates the following: the Coriolis force is proportional to the velocity relative to the earth; if there is no velocity, there is no Coriolis force. The Coriolis force increases with increasing latitude; it is a maximum at the North and South poles and is zero at the equator. The Coriolis force always acts at right angles to the direction of motion. In the Northern Hemisphere it acts to the right (for an observer looking in the direction of motion); in the Southern Hemisphere (where the sine of the latitude is negative) the Coriolis force acts to the left.

In a system in which the Coriolis force is important, physical intuition is of little help in predicting what will occur. Pendulums rotate and particles are accelerated normal to their direction of movement. An interesting example is to consider what happens to a ball that rolls down a frictionless inclined plane with a slope i. In the absence of a Coriolis term, the governing equation is simply

$$\frac{du}{dt} = -gi \tag{5.15}$$

and, assuming that the ball starts from a resting position at the top of the incline, the velocity at time t is given by

$$u = -git \tag{5.16}$$

and the distance traveled by

$$X = -\frac{1}{2} git^2 \tag{5.17}$$

However, if the Coriolis force is added, the equation becomes

$$\frac{du}{dt} = -gi + fv$$
$$\frac{dv}{dt} = -fu \tag{5.18}$$

The solution of this is

$$X = -\frac{gi}{f^2}(1 - \cos ft)$$

(5.19)

$$Y = +\frac{gi}{f^2}(ft - \sin ft)$$

The path described by the ball is indicated in Figure 5.7. The ball starts down the inclined plane, but as soon as it begins to move, the Coriolis force begins to accelerate it to the right, *normal to the incline* As the ball picks up speed, the Coriolis effect becomes larger and the curvature becomes noticeable. Eventually, the ball is running normal to the incline. However, the Coriolis force continues to accelerate the ball to the right, and now the ball starts to run back up the incline. As it does, it slows down. Under the assumption of no frictional losses, the ball will continue to curve up the inclined plane, continuously losing speed until it reaches the top. At that point the velocity is zero (and so, therefore, is the Coriolis force). The ball now begins once more to roll down the inclined plane, and the process is repeated.

It is doubtful if anyone will ever measure this type of motion in a laboratory. The ball must roll for 5 min to see a curvature of 1 part in 100. Even with an incline of only 0.1%, the ball will have traveled nearly 500 m in this time, and for the ball to reach the bottom and

Figure 5.7. Under the assumptions of the geostrophic equation for a flat earth, a particle on an inclined plane follows the path of a cycloid (see text for argument).

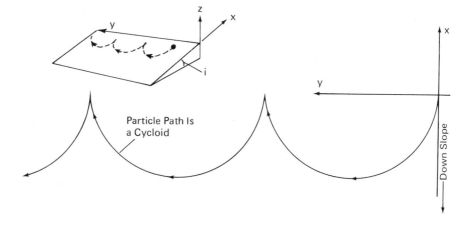

Particle Path Is
a Cycloid

Down Slope

come back up to the top would require an inclined plane somewhat larger than the United States. To do the experiment on an inclined plane of only a few square miles [where the flat-earth assumptions implied in Eq. (5.21) might be expected to hold] would require an inclined plane whose slope was measured in parts per million.

Gravity

With the assumed coordinate system, gravity acts along the z axis. Although gravity varies slightly from place to place, the change is insignificant for any problem in physical oceanography. Surface gravity changes about 0.5% (978 cm/s² at the equator and 983 cm/s² at the poles). The decrease in gravitational potential is related to the spinning earth. The first term in Eq. (5.14) is centrifugal acceleration, which varies between zero at the poles and 3.4 cm/s² at the equator. (See also equation I-15 and I-17 in Appendix I.) The remainder of the 5-cm/s² difference between the poles and the equator is related to the fact that the radius of the earth at the equator is about 22 km larger than the polar radius, a circumstance that can be explained in part by the expected equilibrium shape of a rotating earth.

If the earth were of uniform density, gravity would decrease linearly with depth; but because the earth's density increases with depth, gravity actually increases with depth through the crust. The change is small:

$$g(z) = g_0 + 2.3 \times 10^{-4}Z \text{ cm/s}^2 \tag{5.20}$$

where the depth is measured in meters. Even in the bottom of the deepest trench, the value of gravity is only 0.25% larger than at the surface.

Gravity measurements at sea are usually made from a moving ship. An instrument that measures acceleration cannot distinguish one type of acceleration from another. Short-period accelerations, such as the roll and heave of the ship, can be averaged out, but the centrifugal acceleration due to the ship's movement cannot. In Eq. (5.14) and Figure 5.6 the term $2\Omega u$ was divided into horizontal and vertical components. The horizontal component was the Coriolis force; the vertical component, $2\Omega u \cos \phi$, points in the direction of the gravitational vector and is the Eötvös correction, which must be applied to all observations of gravity if made from a moving platform. A ship with an eastward velocity of 10 knots would have an Eötvös correction of at least 50 milligals (1 gal = 1 cm/s^2) in mid-latitudes.

Furthermore, since we have adopted a right-handed coordinate system, the z axis is up, and gravity should have a negative value since it acts toward the center of the earth. Similarly, all ocean depths should have negative values, and these should become increasingly negative as the depth increases. The convention is ignored in this section and throughout the text where possible. Ocean depths are treated as positive values. The reader, however, is warned to be wary of this problem if he (or she) should attempt analytical solutions of equations similar to Eq. (5.26).

Friction

The fourth and final force to be discussed is friction. A wind blowing on the surface of the water will set the surface water in motion. Because the water is viscous, the frictional stress applied to the water will be transmitted downward. If the wind stops, the water will begin to slow down, and eventually the movement will stop as the effect of water viscosity acts to transfer the kinetic energy to heat energy.

The molecular viscosity of water is known, and the rate of energy transfer and dissipation can be calculated. According to such calculations, the effect of a 20-knot wind blowing on the surface of the water for 48 h will be barely discernible 2 m below the surface. Even the most casual observer knows that the effect of such a wind will be felt much deeper. The problem is not with the molecular theory of viscosity, but rather that the stresses are not transferred by molecular processes alone. A similar problem occurs in predicting the rate at which a fluid slows down once the wind stress has been removed. The slowing down of a fluid in motion occurs many times faster than molecular theory predicts.

One way to visualize the problem is to consider how momentum is transferred by molecular processes. Consider a surface stress that imparts a velocity of 1 knot at the surface. Momentum is transferred downward by molecular motion, the movement of molecules associated with the thermal energy of the fluid. As these molecules move randomly through the fluid, the microscopic flow pattern gradually changes.

Assuming that there were some way to maintain a constant velocity at the surface, the flow after 1 day, 10 days, and 1 year would be as in Figure 5.8, where the dimensions and scales are correct for water with a molecular viscosity of 10^{-2} g/cm s. It is not easy to

Figure 5.8. Velocity distribution after 1 day, 10 days, and 1 year based on molecular viscosity, assuming a 1-knot surface current.

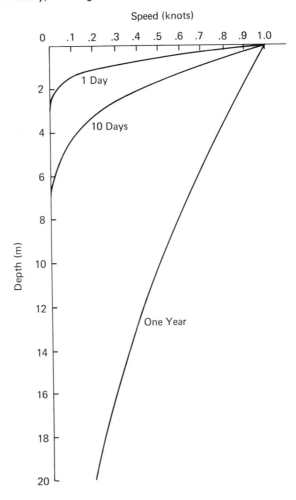

perform such an experiment in the laboratory even for periods of a day, let alone a year. Unless one is careful in the design, the flow will not be laminar, but turbulent.

All motion is turbulent in the ocean. Rather then having momentum transferred downward by molecular motion alone, there is also transfer by turbulent motion. Some have attempted to consider the turbulent transfer process as analogous to the molecular, where globs or eddies of moving fluid are analogous to molecules, and "mixing lengths" are equivalent to molecular mean free path lengths. A coefficient of eddy viscosity many times larger than the coefficient of molecular viscosity can be substituted. However, there is a problem of scale when such a substitution is made, whether or not one believes in the validity of the analogy. For example, vertical mixing, resulting from globs of water a few centimeters in size with a characteristic mixing length of 1 m, would be much less than that resulting from larger eddies 200 km across, which break off from the Gulf Stream. Thus a wide variety of values for eddy viscosities can be found in the oceanographic literature.

Frictional processes dissipate kinetic energy as well as redistribute it. Ultimately, the transfer of kinetic energy to heat energy must be by molecular viscosity processes. But, as in the problem of energy redistribution by wind-induced shearing stress, a consideration of molecular processes and molecular coefficients is insufficient to account for observations. A few measurements suggest that the transfer of kinetic energy to heat energy is at the rate of 10^{-4}–10^{-5} erg g^{-1} s^{-1}. There is some suggestion that the transfer is higher in regions of high current shear. Values of 10^{-2} have been found in strong tidal currents and in the region of the thermocline.

If eddy viscosity is considered an exact physical analogy to molecular viscosity, the frictional terms in the x, y, and z directions are (see Appendix I)

$$\text{friction } (x) = \frac{A_h}{\rho}\left(\frac{\partial^2 u}{\partial x^2} + \frac{\partial^2 u}{\partial y^2}\right) + \frac{A_z}{\rho}\frac{\partial^2 u}{\partial z^2}$$

$$\text{friction } (y) = \frac{A_h}{\rho}\left(\frac{\partial^2 v}{\partial x^2} + \frac{\partial^2 v}{\partial y^2}\right) + \frac{A_z}{\rho}\frac{\partial^2 v}{\partial z^2} \tag{5.21}$$

$$\text{friction } (z) = \frac{A_h}{\rho}\left(\frac{\partial^2 w}{\partial x^2} + \frac{\partial^2 w}{\partial y^2}\right) + \frac{A_z}{\rho}\frac{\partial^2 w}{\partial z^2}$$

where the subscripts h and z denote horizontal and vertical eddy viscosity coefficients. Characteristic values are

$A_h = 10^4$–10^7g/cm s

$A_z = 1$–10 g/cm s

The effect of the wind blowing on the surface of the ocean and the consequent transfer of momentum to the ocean by eddy viscosity can be written

$$\frac{\partial \tau_x}{\partial z} = A_z \frac{\partial^2 u}{\partial z^2}$$

$$\frac{\partial \tau_y}{\partial z} = A_z \frac{\partial^2 v}{\partial z^2}$$

(5.22)

where τ_x and τ_y are the x and y components of wind stress.

Perhaps the simplest form of the frictional terms is to simply set friction proportional to velocity:

$$\text{friction } (x) = -Ju$$

$$\text{friction } (y) = -Jv$$

(5.23)

$$\text{friction } (z) = -Jw$$

None of these forms are very useful in thinking about the physics of turbulent transfer processes or of turbulent dissipation. A more useful way is to consider currents to be composed of two components, a mean or average component and a turbulent component. The flow recorded by a current meter for 1 h can be averaged to give the mean flow $\bar{u}$ If this mean flow is subtracted from the instantaneous current, the remainder may be termed the turbulent component, u'. Thus the instantaneous current u can be written as the sum of the mean flow $\bar{u}$ and the turbulent flow u':

$$u = \bar{u} + u'$$

(5.24)

This separation can be carried out within the various terms in Eq.

(5.5) in such a way that Eq. (5.5) can now be written in terms of the mean flow plus a series of remainder terms, which includes u', v', w'. These frictional terms are called Reynolds stresses (see Appendix I). They are

$$\text{friction } (x) = -\frac{\partial}{\partial x} \overline{(u'u')} - \frac{\partial}{\partial y} \overline{(u'v')} - \frac{\partial}{\partial z} \overline{(u'w')}$$

$$\text{friction } (y) = -\frac{\partial}{\partial x} \overline{(u'v')} - \frac{\partial}{\partial y} \overline{(v'v')} - \frac{\partial}{\partial z} \overline{(v'w')} \qquad (5.25)$$

$$\text{friction } (z) = -\frac{\partial}{\partial x} \overline{(u'w')} - \frac{\partial}{\partial y} \overline{(v'w')} - \frac{\partial}{\partial z} \overline{(w'w')}$$

Equations of Motion

To summarize briefly: the forces that balance the acceleration term are four in number, (1) the pressure-gradient force, (2) gravity, (3) friction, and (4) the Coriolis force; the latter is in some sense a fictitious force in that it is required because we assume a nonrotating coordinate system when in fact all our observations are made on a rotating earth. We have further defined our coordinate system so that the gravity vector points in the vertical direction, which eliminates the gravity term from the xy plane. Finally, we note that there are several choices for the analytical expressions for the frictional terms, none of which is particularly satisfactory. For purposes of this text, we chose the simplest.

It is now possible to write Eq. (5.5) analytically. One form is

$$\frac{du}{dt} = -\frac{1}{\rho}\frac{\partial p}{\partial x} + fv + \frac{1}{\rho}\frac{\partial \tau_x}{\partial z} - Ju$$

$$\frac{dv}{dt} = -\frac{1}{\rho}\frac{\partial p}{\partial y} - fu + \frac{1}{\rho}\frac{\partial \tau_y}{\partial z} - Jv \qquad (5.26)$$

$$\frac{dw}{dt} = -\frac{1}{\rho}\frac{\partial p}{\partial z} - g - Jw$$

The τ term refers to the stress of the wind on the ocean surface.

Internal frictional losses are assumed proportional to velocity. Gravity includes the effect of centrifugal acceleration. Second-order Coriolis terms have been ignored.

It is important to write these equations down and to note the individual terms, not because there will be any attempt in this text to consider Eq. (5.26) in any detail, but because the relationships that are discussed in the next two chapters should be considered against Eq. (5.26).

chapter 6

BALANCE OF FORCES: WITHOUT CORIOLIS FORCE

The equations of motion described in Chapter 5 are simply the fluid dynamic, or oceanic, eqivalent of Newton's second law, which states that the acceleration of a particle is proportional to the forces applied. It cannot be emphasized too strongly that there are no known analytical solutions to the equation represented by Eq. (5.26). We deal with various simplifications and approximations of these equations. In a number of cases we can be quite confident of solutions based on approximations and the ignoring of certain terms. At times, however, our solutions bear only a distant resemblance to what we observe in the ocean. However, even in these cases, the simplified mathematical model often provides insight into the physical processes occurring. In this chapter we consider some simple balance relationships that ignore the Coriolis force, but which are of some relevance to the ocean. In Chapter 7 we consider force balances in which the Coriolis force plays a predominant role.

Hydrostatic Equation

The hydrostatic equation relates the static pressure within a fluid to the overlying weight of this fluid. The pressure at some depth $-Z$ is simply

$$p = - \int_0^{-Z} \rho g \; dz \qquad (6.1)$$

and in the case where density is constant

$$p = + \rho g Z \qquad (6.2)$$

The hydrostatic equation can be considered a special case of the vertical component of the equation of motion, Eq. (5.26), where there is no friction or vertical acceleration. In this case Eq. (5.26) reduces to

$$\frac{1}{\rho} \frac{\partial p}{\partial z} = -g \qquad (6.3)$$

which can be integrated to give Eq. (6.1).

The hydrostatic approximation is a quite general solution to the vertical component of Eq. (5.26). One case where it is not is when one considers wave motion. In that case the vertical component of acceleration must be included (see Chapter 10 and Appendix I).

Since the density of seawater is slightly greater than unity, and the acceleration of gravity is slightly less than 1000 cm/s², the pressure of 10 m of seawater is very close to 1 bar (1 million dynes/cm²), which is the approximate pressure of the standard atmosphere. Thus every increase in depth of 10 m is the equivalent of increasing the pressure about 1 atmosphere.

Because of this simple numerical relationship, oceanic pressure is often measured in decibars (one tenth of a bar), since 1 decibar is approximately equivalent to 1 m in depth. Most "depth measurements" within the water column are measured by pressure-recording devices of one kind or another. Usually, these are calibrated in terms of 1 m equals 1 decibar. Although the agreement deteriorates slightly at great depths as compressibility increases the density of seawater, the differences are relatively small. The pressure

gauge on the bathyscaphe *Trieste* on her dive to the bottom of the
Marianas Trench recorded a pressure of 11,240 decibars. Systematic
soundings in the area indicated a maximum depth of about 10,880 m
for the trench.

The hydrostatic equation can be used in simplifying the
pressure-gradient term in Eq. (5.26). Consider first a fluid of constant
density. If the surface of the fluid were level, there would be no
horizontal pressure gradient; but if the fluid surface sloped, there
would be a horizontal pressure gradient.

$$\frac{1}{\rho}\frac{\partial p}{\partial x} = \frac{1}{\rho}\frac{p_b - p_a}{\Delta X}$$

$$= \frac{1}{\rho \Delta X}[\rho g(Z + \Delta Z) - \rho g Z] \tag{6.4}$$

$$= g\frac{\Delta Z}{\Delta X}$$

$$= g i_1$$

where i_1 is the slope of the surface in the x direction (Figure 6.1a). It is
easy to demonstrate that this pressure gradient applies through the
entire fluid.

Consider next a two-layer fluid. The upper surface is level, but
there is a slope to the interface (Figure 6.1b). Then the pressure-
gradient term in the bottom layer becomes

$$\frac{1}{\rho}\frac{\partial p}{\partial x} = \frac{1}{\rho_2}\frac{p_b - p_a}{\Delta X}$$

$$= \frac{1}{\rho_2 \Delta X}[(\rho_1 g Z_1 + \rho_2 g \Delta Z + \rho_2 g Z_2) - (\rho_1 g Z_1$$
$$+ \rho_1 g \Delta Z + \rho_2 g Z_2)] \tag{6.5}$$

$$= g\frac{\rho_2 - \rho_1}{\rho_2}\frac{\Delta Z}{\Delta X}$$

$$= g\left(\frac{\rho_2 - \rho_1}{\rho_2}\right) i_2$$

where i_2 is now the slope of the interface. There is no horizontal
pressure gradient in the upper layer, and the pressure gradient calculated above applies throughout the lower layer.

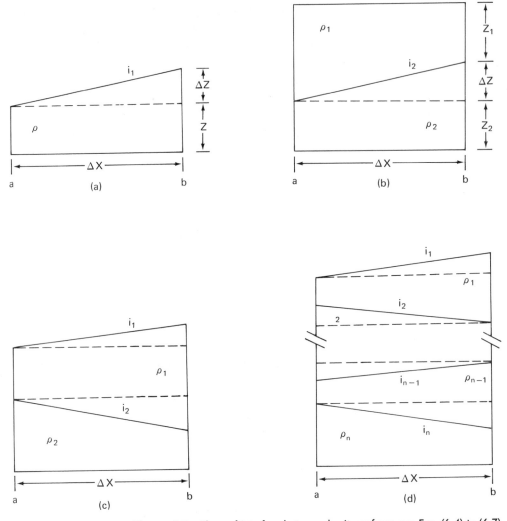

Figure 6.1. Slope of interface between density surfaces; see Eqs. (6.4) to (6.7).

If there is a slope to the sea surface (Figure 6.1c), it is fairly easy to show that the horizontal pressure gradient in the bottom layer is simply

$$\frac{1}{\rho_2}\frac{\partial p_2}{\partial x} = g\left(\frac{\rho_2 - \rho_1}{\rho_2}\right) i_2 + \frac{1}{\rho_1}\frac{\partial p_1}{\partial x} \tag{6.6}$$

and by a simple extension of this logic, the horizontal pressure gradient in any layer (n) is related to the pressure gradient in the layer above it (Figure 6.1d) by the relationship

$$\frac{1}{\rho_n}\frac{\partial p_n}{\partial x} = g\left(\frac{\rho_n - \rho_{n-1}}{\rho_n}\right)i_n + \frac{1}{\rho_{n-1}}\frac{\partial p_{n-1}}{\partial x} \tag{6.7}$$

This relationship is the basis from which Margule's equation is derived (see Chapter 7).

Dynamic Height

In oceanography the hydrostatic equation is often written

$$\alpha \, dp = -g \, dz \tag{6.8}$$

where α is the specific volume. Equation (6.8) has units of energy per unit mass. In oceanography the integral of Eq. (6.8) over a given pressure interval is called *dynamic height D*.

$$D = \int_{p_1}^{p_2}\alpha \, dp$$
$$dD = \alpha \, dp \tag{6.9}$$

Measuring the differences in dynamic height between two pressure surfaces is accomplished by carefully measuring the density (specific volume) distribution within the water column. As described in Chapter 1, density in situ is not measured directly, but rather the temperature and salinity of the water column are measured as a function of pressure (depth). The equation of state of seawater is known sufficiently well that density can be calculated to a few parts in a hundred thousand from measurements of temperature, salinity, and pressure. The dynamic height between two pressure surfaces, Eq. (6.9), can be found from the measured values by numerical integration. Because of their importance in determining geostrophic currents (see Chapter 7), these techniques have reached a high degree of refinement in the past half-century.

Comparing differences in dynamic heights between adjacent stations *a* and *b* is the equivalent of measuring the differences in the horizontal pressure gradient over the same depth (pressure) interval between stations a and b:

$$D_a = \int_{p_1}^{p_2} \alpha_a \, dp$$

$$D_b = \int_{p_1}^{p_2} \alpha_b \, dp \tag{6.10}$$

$$D_a - D_b = \int_{p_1}^{p_2} (\alpha_a - \alpha_b) dp$$

The usual practice is to subtract out a "standard ocean" and to measure something called *dynamic height anomaly* ΔD:

$$\alpha = \alpha_{35,0,p} + \delta \tag{6.11}$$

$$\Delta D_a - \Delta D_b = \int_{p_1}^{p_2} (\delta_a - \delta_b) dp \tag{6.12}$$

In terms of the pressure-gradient term in Eq. (5.26),

$$\frac{1}{\rho} \frac{\partial p}{\partial x} = \frac{\partial(\Delta D)}{\partial x} \tag{6.13}$$

River Flow

As an introduction to estuarine circulation, and as a contrast to the balance of forces with the Coriolis term included to be considered in Chapter 7, consider the simple problems of balance of flow in a river. Let us assume that the river runs in an east–west direction so we can ignore the y component of Eq. (5.26); and let us further assume that wind stress and the Coriolis force can be ignored, and that the flow is steady state and uniform (i.e., acceleration is zero). The balance of forces is simply the hydrostatic equation for the vertical and a balance between the pressure-gradient term and friction in the horizontal:

$$\frac{1}{\rho} \frac{\partial p}{\partial x} = -Ju \tag{6.14}$$

$$\frac{1}{\rho}\frac{\partial p}{\partial z} = -g \tag{6.15}$$

And since we can assume uniform density in the river, we can write the pressure-gradient term in terms of the surface slope:

$$gi_x = -Ju \tag{6.16}$$

If the slope of the Mississippi were about 1 in 10,000, and if you assume a value of 10^{-3}/s for the internal friction term J, then the velocity of the river is about 100 cm/s, or approximately 2 knots:

$$u = -\frac{g}{J}i_x \cong -\frac{10^3}{10^{-3}} \times 10^{-4} = -10^2 \tag{6.17}$$

If you increase the slope, the current speeds up. If you increase the frictional term (such as by making the river shallower or narrower without changing the slope of the river surface), you decrease the speed. In this example, water flows downhill as expected.

Convective Flow in an Estuary

Consider next a river that does not flow over a solid river bed, but over a layer of saltwater, and into the ocean (Figure 6.2a, b). Such is the basic requirement for an *estuary,* which is often defined as a semienclosed body of water having a free connection with the open sea and within which the seawater is measurably diluted with freshwater deriving from land drainage. If there were no turbulent mixing between the river and the underlying saltwater, the balance equations would be those for an ordinary river situation as described in the last section. The river water would simply slip over the salt wedge. However, turbulent mixing does occur; some saltwater is mixed upward and freshwater downward. As a result of this mixing, the salinity of the inner edge of the salt wedge is reduced, which reduces the density of the salt wedge, which in turn establishes a horizontal pressure gradient to drive the salt wedge farther up the river against the current. This in turn increases the shear between the river and the salt wedge, which enhances turbulent mixing, resulting in an increased horizontal pressure gradient and more inward flow (Figure 6.2b).

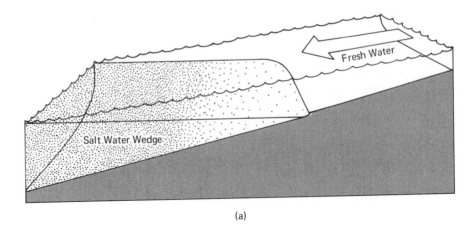

(a)

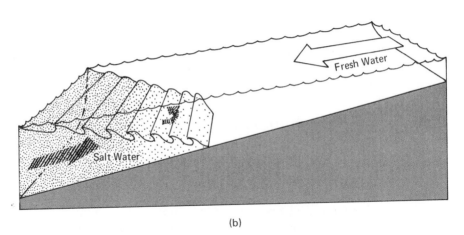

(b)

Figure 6.2. (a) In the absence of mixing, freshwater would flow over a saltwater wedge and spread out over the surface of the ocean; (b) because of mixing, some of the saltwater is mixed upward into the freshwater layer.

In an actual estuary, the mixing between the river water and the underlying salt wedge is generally greatly enhanced by mixing caused by tidal currents, but the principle is the same. One consequence of the mixing of fresh- and saltwater is that there is a net inflow at the bottom of an estuary and a net outflow at the surface, which can be several times the size of the river outflow before it enters the estuary. The relationship can be shown qualitatively using the simple conser-

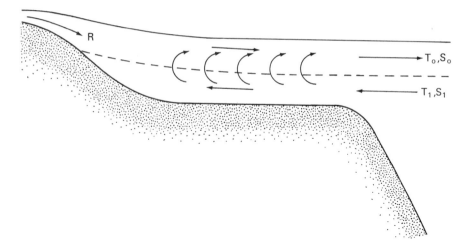

Figure 6.3. Relationship between river flow, mixing, and transport in and out at the mouth of an estuary; see Eq. (6.18).

vation arguments of Chapter 4. Let the volume transport and the mean salinity in the upper and lower section of the estuary be as shown in Figure 6.3. If the volume transport of the river is R, Eqs. (4.20) and (4.21) can be written as

$$\left. \begin{array}{l} T_0 = T_i + R \\ T_0 S_0 = T_i S_i \end{array} \right\} \quad (6.18)$$

Solving for T_i and T_0 gives

$$\left. \begin{array}{l} T_i = \dfrac{S_0}{S_i - S_0} R \\[2ex] T_0 = \dfrac{S_i}{S_i - S_0} R \end{array} \right\} \quad (6.19)$$

With the river flow constant, a decrease in stratification (i.e., more vertical mixing and a weaker salinity gradient) increases the transport in the upper and lower layers. In other words, more vertical mixing results in more water being transported seaward in the upper layer; consequently, more water must be transported landward in the bottom layer to conserve the mass of salt and water in the estuary.

An increase in river flow can increase the shear and thus also increase the mixing. Many years ago engineers diverted a second river to flow into a major estuary with the idea that the increased flow would keep the channel from silting up and thus reduce the cost of dredging. The result was an increase in the flow *up* the estuary and a consequent increase in the rate of silting and cost of dredging.

Conservation arguments are important tools for describing net estuarine flow; however, they do not provide insight into the actual driving mechanism. For this, the dynamics of net estuarine circulation must be discussed. The increased flow up the estuary along the bottom is maintained at least in part by gravitational forces. In the surface layer the sea surface slopes down to seaward, resulting in a pressure gradient that drives the water of the surface layer out of the estuary. The reversal in pressure gradient with depth can result from the slope of the interface between the two layers, which is in the opposite sense of the surface layer, and/or from the upstream decrease in the average density of the water column resulting from vertical mixing. This pressure force generates a flow up the estuary in the bottom layer (Figure 6.4).

Types of Estuaries: Estuaries can be usefully categorized by their degree of vertical stratification. They vary from highly stratified estuaries to those that are well mixed and with little or no vertical salinity gradient (Figure 6.5). Deep fjords, with or without a shallow sill, are usually highly stratified (type C in Figure 6.5), with freshwater flowing out over a nearly uniform deep saline layer. A second type of highly stratified estuary is the *salt wedge* (type D in Figure 6.5), which occurs where the ratio of parameters relating river outflow to vertical mixing by tides is very large. The Mississippi River is an example. As the ratio of river flow to tidal mixing parameters decreases, the estuary becomes less stratified. The

Figure 6.4. (a) Downstream increase in salinity results in a downstream increase in density. If the slope of the sea surface were level, the pressure-gradient force would result in a flow upstream. (b) Because the free surface slopes to seaward, there is a reversal in the pressure gradient, which results in seaward flow at the surface and landward flow at depth. Note that the isobaric surfaces apply to the net or time average motion. Tidal motion and wind-driven currents are superimposed on this net flow.

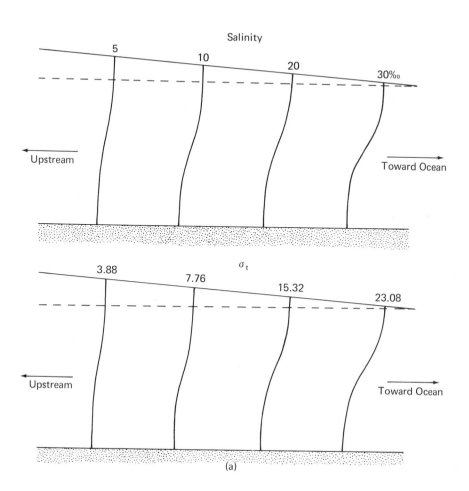

Salinity

Upstream ← → Toward Ocean

σ_t

Upstream ← → Toward Ocean

(a)

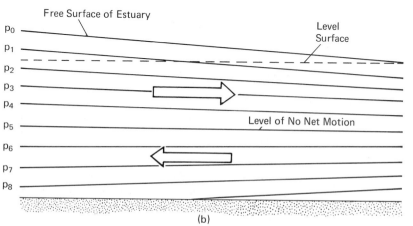

Free Surface of Estuary

p_0

p_1

p_2

p_3

p_4

p_5

p_6

p_7

p_8

Level Surface

Level of No Net Motion

(b)

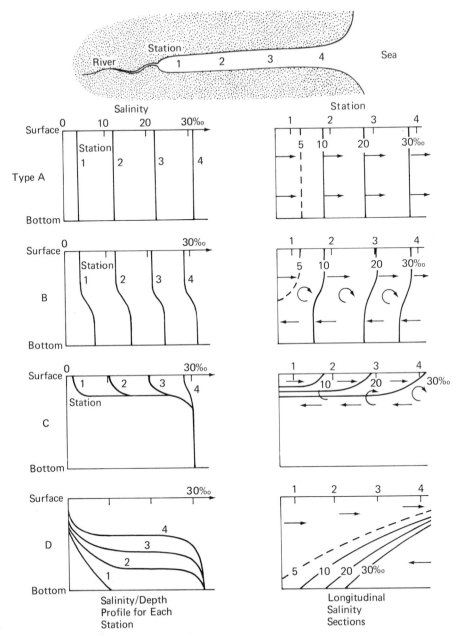

Figure 6.5. Longitudinal salinity distribution and resulting net flow patterns for (a) a well-mixed estuary, (b) a partially mixed estuary, (c) a fjord-type estuary, and (d) a salt-wedge estuary.

James River in the Chesapeake Bay is perhaps the best studied example of a slightly stratified estuary (type B in Figure 6.5). All the major coastal plain estuaries of the U.S. east coast, such as Narragansett Bay, Delaware Bay, and Chesapeake Bay, are examples of partially mixed estuaries. The vertical salinity gradient, however, can be very small, as is the case in Narragansett Bay.

The extreme is the case where vertical mixing is so strong compared to the river runoff that there is no vertical salinity gradient (type A in Figure 6.5). The salinity increases from head to mouth. Shallow estuaries with strong tides can lead to well-mixed estuaries. The Severn River in England is an example.

Two criteria must be satisfied for a salt-wedge estuary to exist. The first is that the ratio of tidal flow to river flow be small (less than unity); that is, the volume of river water entering the estuary between low and high tide must be larger than the volume of seawater entering the estuary over the same time span. Second, the width of the estuary must not increase significantly over that of the river. The lower portion of the Mississippi River is an example of a salt-wedge estuary.

As tidal mixing and width increase or as river input and depth decrease the degree of stratification also decreases, and the estuary becomes partially mixed (type B in Figure 6.5). Most estuaries fall under this category. The ratio of tidal flow to river flow for a partially mixed estuary lies between 10 and 10^3. For Narragansett Bay, this ratio is of the order of 10^2.

Tidal mixing may be strong enough to all but eliminate vertical stratification. This may occur when the ratio of tidal to river input is greater than 10^3. The result has been termed a well-mixed, or vertically homogeneous, estuary. The forces that determine the flow in a completely vertically mixed estuary are more difficult to explain. Two types of vertically homogeneous estuaries have been discussed in the literature: (1) the sectionally homogeneous estuary, and (2) estuaries exhibiting lateral variations in salinity. Gravitational convection weakens as an estuary approaches sectional homogeneity. A truly sectional homogeneous estuary with a mean seaward flow at all depths cannot exist.

It is natural to break the discussion of the water motion in an estuary into two distinct (but related) topics, (1) instantaneous motion, and (2) net motion. Instantaneous motion is characterized by the ebb and flood of the tides at all depths. Upon averaging out the tidal currents over one or more tidal cycles, a net current remains that varies with depth. The instantaneous motion is often an order of magnitude larger than the net motion, as, for example, in the West Passage of Narragansett Bay, where typical tidal current amplitudes are 20–50 cm/s, whereas typical net speeds are about 5–10cm/s (Figure 6.6). Although small, the net motion is important since it actually transports water and other matter into and out of the estuary, whereas tidal currents essentially move water to and fro without resulting in a net displacement. One consequence of the differential net flow with depth of a partially mixed estuary is the phase shift in the tidal currents. The surface ebb tide starts sooner and flows longer than that below the halocline (Figure 6.7).

It must be recalled that gravitational convection is not the only mechanism driving the net circulation. River runoff, atmospheric pressure, and winds are also factors. Normal variations in river runoff do not have a marked effect on the net circulation because the stratification readjusts itself in opposition to the change. The inverse barometer effect of atmospheric pressure is also generally unimportant except in cases of large pressure variations, such as those that accompany storm surges. The effects of everyday wind fluctuations, however, are *very* important. The net circulation will tend to be increased or decreased in response to the wind stress on the estuarine surface. For this reason, estuarine measurements made during different wind conditions may show marked variations (see Figure 6.6).

Finally, it should be noted that the effects of the earth's rotation cannot be completely ignored (see Chapter 7). In broad estuaries there is a tendency to form a cross-stream slope at the surface and the interface, as shown in Figure 6.8. Likewise, there is a tendency for the outflow to hug the right side (facing downstream) of the estuary in the Northern Hemisphere, which sometimes results in a horizontal salinity gradient (fresher water on the right). There is likely to be a stronger ebb and weaker flood flow on the right side of the estuary. In the absence of strong coastal currents, there is also a tendency for the outflow of rivers and estuaries to curve to the right in the Northern Hemisphere once the outflow reaches the open ocean.

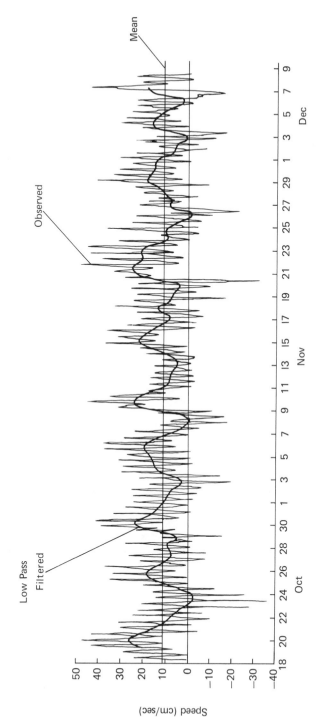

Figure 6.6. Instantaneous velocity is generally dominated by the tides. In this example for Narragansett Bay, the tidal range is of the order of 40 cm/s. Superimposed on the tides is a slowly varying mean velocity (noted as "low pass filtered"), which is primarily controlled by changing wind conditions that redistribute the water in the bay and hence affect the pressure field. The mean flow over the 3-month interval is 11 cm/s and is due essentially to the gravitational-convective flow described in Figure 6.4. (After Weisberg, R. H., 1976: "A Note on Estuarine Mean Flow Estimation," *J. Mar. Res.*, 34.)

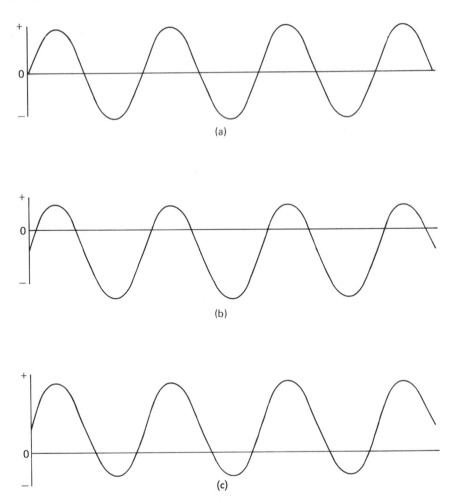

Figure 6.7. (a) Simple idealized sinusoidal 12.5-h semidiurnal tidal flow in the absence of any net flow; (b) and (c) same flow with a net seaward current in the upper layer (b) and a net landward current in the bottom layer (c). As a result the ebb tide lasts longer than the flood tide in (b), and the flood longer than the ebb in (c). There is also a shift in the phase. For example, the ebb tide starts at the surface sooner than in the bottom layer.

Terminal Velocity

Imagine a small clay particle and a spherical equipment capsule falling through the ocean. If they were to fall in a vacuum they would continuously accelerate at 9.8 m/sec² and would fall at the same

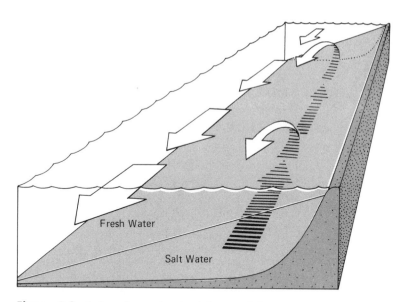

Figure 6.8. In broad estuaries the influence of the Coriolis force cannot be ignored. One result is a slope of the interface as shown. The slope would be reversed in the Southern Hemisphere.

speed. In the ocean the clay particle and the sphere soon reach some terminal velocity and continue to fall at that rate; the sphere considerably faster than the clay particle even if the density of the two are identical.

Both are examples of steady-state force balances where the downward gravitational forces acting on the object are balanced by an equal and opposite frictional force.

$$mg = F \tag{6.20}$$

In the case of the clay particle, the frictional force is proportional to the molecular viscosity while for the sphere the forces are controlled by the turbulent wake left by the falling sphere. In both cases the gravitational force mg is the weight of the object in the water

$$mg = (\rho_s - \rho_w) \, Vg \tag{6.21}$$

where ρ_s is the density of the object and ρ_w is the density of water and V the volume of the object. For a sphere it is

$$mg = \frac{4}{3} \pi r^3 \left(\rho_s - \rho_w\right)g \tag{6.22}$$

The frictional viscous drag of the spherical clay particle as given by Stokes is

$$F = 6 \pi r \mu w \tag{6.23}$$

which combined with Eq. (6.22) yields

$$w = \frac{2}{9} r^2 \frac{g}{\mu} (\rho_s - \rho_w) \tag{6.24}$$

Equation (6.24) holds for Reynolds numbers less than 1, where the nondimensional Reynolds number is defined as

$$R_e = \frac{2wr\rho_w}{\mu} \tag{6.25}$$

For reasonable values of $\rho_s - \rho_w$ and a value of $\mu \cong 10^{-2}$ g/cm s for water, it is easy to show that Eq. (6.24) holds for objects with a radius no greater than a few tens of microns.

For an object whose radius is as great as 10^{-2} cm, the frictional drag caused by molecular viscosity is no longer controlling. The frictional force is given by

$$F = \tfrac{1}{2} C_D \rho_w B w^2 \tag{6.26}$$

where B is the cross-sectional area (πr^2 for the case of a sphere), and C_D is a nondimensional constant whose value is approximately unity. It can be as small as 0.1 for highly streamlined objects and almost 2 for hollow hemispheres (such as nonporous parachutes). For a sphere, it is about 0.5. Combining Eq. (6.26) for a sphere with Eq. (6.22) gives

$$w^2 = \frac{16}{3} \frac{\rho_s - \rho_w}{\rho_w} gr \tag{6.27}$$

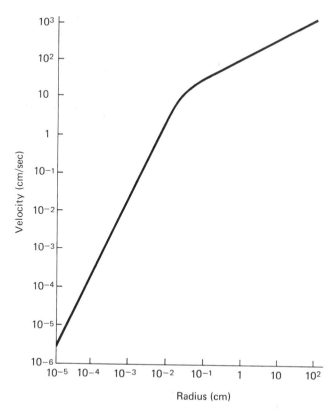

Figure 6.9. Terminal velocity for falling spheres in the ocean as a function of radius. The assumed specific gravity is 2.6. Stokes law applies for objects of the order of 10^{-2} cm and less.

Figure 6.9 gives the terminal velocity for a falling sphere as a function of radius. For the Stokes range, the terminal velocity increases as the square of the radius. For larger objects (and larger falling speeds), the terminal velocity increases as the square root of the radius. A cannon ball of 10-cm radius and specific gravity of 8 will fall at a speed of more than 10 knots and reach a depth of 4000 m in about 11.5 min. A clay particle of 10^{-4} cm and specific gravity of 2.5 will have a terminal velocity of about 3×10^{-4} cm/s and take 40 years to reach the bottom. The relation of sediment distribution patterns on the bottom of the ocean to their sources at the surface suggests that particles reach the

bottom much faster than indicated by the Stokes settling velocity.
Otherwise, ocean currents would transport the sediments many
thousands of miles in the 40 years it would take to fall to the bottom.
It has been suggested that this material is incorporated into the biological cycle and may fall to the bottom as fecal matter or detritus whose
size ranges could be one or two orders of magnitude larger than that
of the clay particle.

chapter 7

BALANCE OF FORCES: WITH CORIOLIS FORCE

In the following sections we examine in some detail the consequences of the Coriolis term to the equation of motion. With geostrophic motion, we balance the Coriolis force against the pressure-
gradient force; with inertial motion, it is Coriolis force against particle
acceleration; with Ekman motion, it is Coriolis force against wind
stress. In the last two sections we looked at some simple examples of
balancing the Coriolis force against more than one term. In each case
the resulting flow is contrary to one's intuitive sense of what should
happen. Even for those with considerable sophistication in physical
concepts, one's first introduction to the consequences of the Coriolis
force often produces something analogous to intellectual trauma.

Geostrophic Flow

Assume that the ocean currents are horizontal and steady, and
that the wind stress and other frictional forces are sufficiently small
that they may be neglected. With the acceleration and friction terms

removed, Eq. (5.26) reduces to the hydrostatic equation in the vertical plane, and a balance between two terms, the pressure force and the Coriolis force, in the horizontal plane.

$$fv = \frac{1}{\rho} \frac{\partial p}{\partial x}$$

$$fu = -\frac{1}{\rho} \frac{\partial p}{\partial y}$$

(7.1)

This is the geostrophic equation, and currents that obey this relation are called *geostrophic currents* All the major currents in the ocean, such as the Gulf Stream, the Antarctic Circumpolar Current, and the equatorial currents are, to a first approximation, geostrophic currents. The consequences of the geostrophic equation are extraordinary. Consider for a moment again the problem of the rolling ball on a frictionless inclined plane. Substituting the slope force for the pressure gradient force,

$$fv = gi_x$$

(7.2)

results in a ball that rolls parallel to the slope (Figure 7.1). No matter what the direction of the slope, the ball rolls parallel to the slope. The geostrophic equation says water does not run downhill; it runs around the hill.

Television has made nearly everyone familiar with weather maps. The winds on weather maps do not blow directly from high-pressure cells to low-pressure cells. The winds' flow is more nearly parallel to the isobars. In the Northern Hemisphere the flow is clockwise around high-pressure cells and counterclockwise around low-pressure cells (Figure 7.1). A good rule of thumb: as you look downstream, the high pressure is on the right in the Northern Hemisphere and on the left in the Southern Hemisphere.

The forces involved in this balance are very small, generally less than $\frac{1}{100}$ dyne/g. Compare this with the hydrostatic force balance, which at 100 m is about 10^7 dynes/g and 5×10^8 dynes/g at 5000 m. But small as they are, the pressure-gradient and Coriolis forces are the largest horizontal forces in much of the ocean. All the major ocean circulation features are in approximate geostrophic balance.

The Gulf Stream flow and all other major surface currents are maintained by the slope of the sea surface. The sea-surface slope

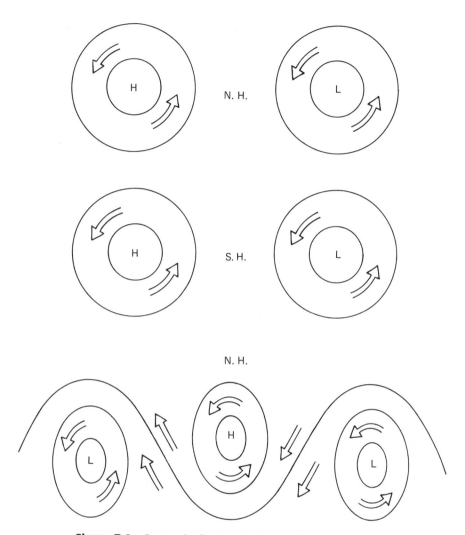

Figure 7.1. Geostrophic flow is *cyclonic* around low-pressure cells and *anticyclonic* around high-pressure cells. In the Northern Hemisphere, cyclonic flow is counterclockwise; it is clockwise in the Southern Hemisphere.

necessary to maintain the Gulf Stream is about 1 part in 100,000. Sea level at Bermuda is about 1 m higher than on the East Coast of the United States as a consequence.

Sea-surface slopes of 10^{-5} and less cannot be measured directly. They are inferred. A variety of techniques have been and are being used. Perhaps the simplest, and also most widely used, is to assume that horizontal pressure gradients decrease with depth, and that if we go deep enough there is no horizontal pressure gradient. There is some evidence to support this hypothesis; much of it is based on the observations that currents generally become weaker as the depth increases, and that the distribution of temperature, salinity, and other properties becomes more uniform with depth. If one evokes the hydrostatic equation or makes the equivalent dynamic-height calculation, the slope of the sea surface can be determined from the density distribution. In the example of Figure 7.2a, the lighter, less dense fluid must stand higher in the tube if there is to be no horizontal pressure gradient at the bottom of the tube where the two fluids meet. The same principle applies in a continuously stratified ocean such as that sketched in Figure 7.2b. If there is no horizontal pressure gradient at depth h, then the pressure at a equals the pressure at b; but if the average density of the water at b is less than at a, then the height of water column b must be greater.

As a real example, consider the temperature distribution across the Gulf Stream (Figure 7.3). The cold, dense coastal water is on the left, the warm, light Sargasso Sea water on the right. Assuming no, or at least a very small, horizontal pressure gradient at a depth of 4000 m implies a thicker water column of lighter warmer water in the Sargasso Sea than in the colder water to the left of the Gulf Stream, which in turn indicates a sea surface that slopes down to the left. The pressure gradient acts to the left and is balanced by the Coriolis force acting to the right, and the Gulf Stream flows in the required direction.

This tendency for the density distribution in the ocean to adjust in such a way as to minimize horizontal pressure gradients at depth allows another rule of thumb: as one faces downstream, the thermocline slopes down to the right (in the Northern Hemisphere); the greater the slope, the stronger the current. What one is really saying,

Figure 7.2. (a) If the two sides of a manometer tube are filled with fluids of different density, and the interface between the two fluids is at the bottom of the tube where there is no horizontal pressure gradient, the lighter fluid will stand higher than the denser fluid. (b) Given the density distribution at stations a and b as shown in the sketch at the upper left, the hydrostatic pressure and the dynamic height are as indicated in the succeeding sketches. If we further assume that at depth h there is a "level surface" (i.e., no horizontal pressure gradient), then the slope to the sea surface and the pressure gradients at depths between h and the surface must be as shown in the lower two sketches.

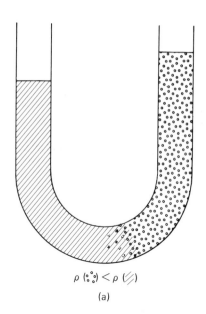

$\rho \, (\substack{\circ \, \circ \\ \circ}) < \rho \, (/\!/)$

(a)

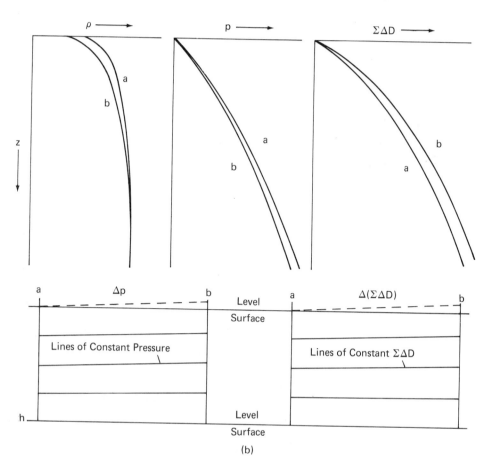

$\rho \longrightarrow$ $p \longrightarrow$ $\Sigma\Delta D \longrightarrow$

z

a

b

a

b

a

b

a Δp b Level a $\Delta(\Sigma\Delta D)$ b

Surface

Lines of Constant Pressure Lines of Constant $\Sigma\Delta D$

h Level

Surface

(b)

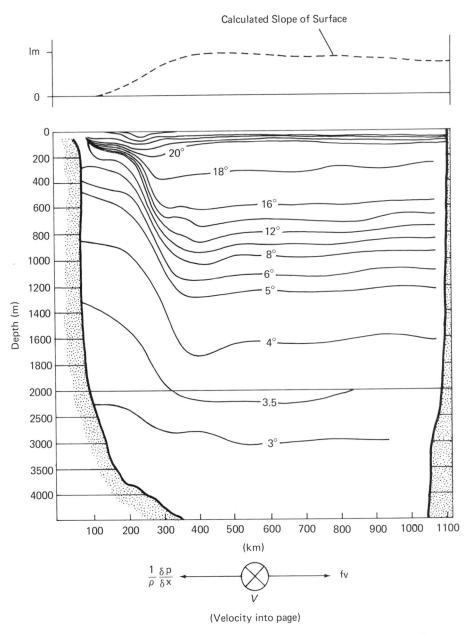

Figure 7.3. Typical temperature distribution across the Gulf Stream. If one assumes no horizontal pressure gradient at 4000 m, then the slope of the sea surface is as indicated. A geostrophic surface current to balance that pressure gradient would be into the paper as shown. (After Iselin, C. O'D., 1936: "A Study of the Circulation of the Western North Atlantic," *Papers in Physical Oceanography and Meteorology*, Vo. 4, No. 4.)

of course, is that as one faces downstream the sea surface slopes up to the right (the sea surface will slope up to the left in the Southern Hemisphere). The greater the slope of the thermocline, the greater the slope of the sea surface and the stronger the surface current. A temperature cross section showing the thermocline and the location of the major equatorial currents in the tropical Pacific provides a good example of the relationship of current direction and thermocline slopes (Figure 7.4). Note that the Pacific South Equatorial Current flows westward along the surface on both sides of the equator, and the thermocline slopes down on both sides of the equator.

Margule's Equation

The ocean is continuously stratified. However, for many purposes it is useful to think of the ocean as comprised of two or more layers, each of constant density. Under such assumptions one can easily calculate the geostrophic flow, knowing the assumed density of each layer, the slope of the sea surface, and the slope (or slopes) of the interfaces between layers. More precisely, Margule's equation allows the calculation of the geostrophic current in the lower of two layers of known density if the geostrophic flow of the upper layer is known as well as the slope of the interface:

$$i_x = +\frac{f}{g} \frac{\rho_2 v_2 - \rho_1 v_1}{\rho_2 - \rho_1} \tag{7.3}$$

when i_x is the slope of the interface in the x direction, and the subscripts 1 and 2 refer to the upper and lower layers, respectively. Equation (7.3) can be applied to any number of successively deeper layers (Figure 7.5). The basis of Margule's equation can be seen by combining Eq. (6.7) with the geostrophic relationship, Eqs. (7.1) and (7.2).

$$\rho_n v_n f = g(\rho_n - \rho_{n-1})i_n + \rho_{n-1}v_{n-1}f \tag{7.4}$$

$$i_n = \frac{f}{g} \frac{\rho_n v_n - \rho_{n-1}v_{n-1}}{\rho_n - \rho_{n-1}}$$

For the special case where the velocity in the bottom layer is zero ($v_n = 0$),

$$i_n = -\frac{f}{g} v_{n-1} \left(\frac{\rho_{n-1}}{\rho_n - \rho_{n-1}} \right)$$

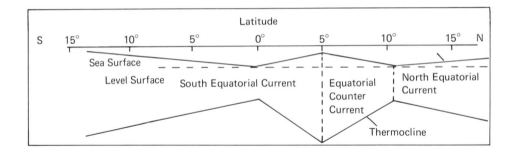

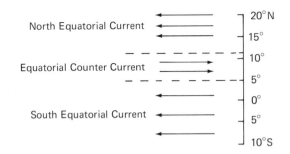

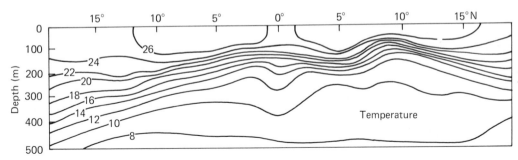

Figure 7.4. North and South Equatorial currents flow westward, separated by the eastward-flowing Equatorial Countercurrent between 5° and 10°N. The currents are mostly confined to the mixed layer about the thermocline. For these currents to be in geostrophic balance, the slope of the thermocline and the sea surface (the latter much exaggerated) are as shown in the top panel sketch; bottom panel shows typical temperature distribution for the North Pacific; the slope of the thermocline generally agrees with the sketch in (a). (After Knauss, J. A., 1963: "Equatorial Current Systems," in *The Sea*, Vol. 2, ed. M. N. Hill, Interscience Publishers, New York.)

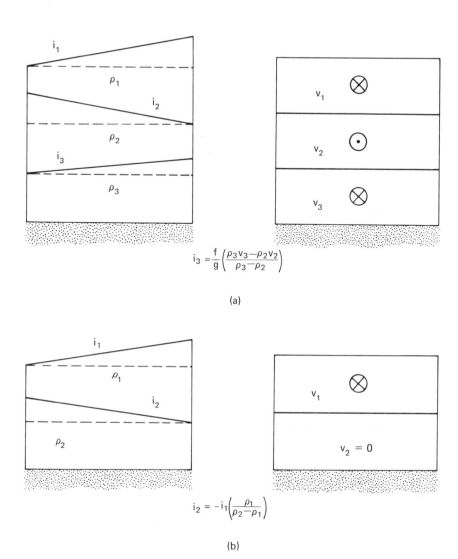

$$i_3 = \frac{f}{g}\left(\frac{\rho_3 v_3 - \rho_2 v_2}{\rho_3 - \rho_2}\right)$$

(a)

$$i_2 = -i_1\left(\frac{\rho_1}{\rho_2 - \rho_1}\right)$$

(b)

Figure 7.5. (a) Relationship among density, geostrophic velocity, and slope of interface between layers as given by Margules equation; (b) the special case of a two-layer ocean with the bottom layer motionless.

or in terms of the slope of the sea surface

$$i_n = -i_{n-1} \frac{\rho_{n-1}}{\rho_n - \rho_{n-1}} \tag{7.6}$$

For a normal thermocline, $\Delta\rho \cong 0.002$ g/s^3, which means that the slope of the thermocline would be approximately 500 times greater than the sea surface, and in the opposite sense.

Wind Stress: Ekman Motion and Upwelling

When the wind blows along the surface of the ocean, it causes both surface currents and waves. The quantitative details of how the stress of the wind is applied to the ocean surface are not very well understood. Energy is transferred by some kind of turbulent process, and fuller understanding requires a detailed examination of not only the mean wind, current, and pressure fields, but of the variations of the wind, current, and pressure about the mean.

Several semiempirical observations are useful. One is that the surface current induced by a wind is approximately 3% of the wind. One might expect a 0.6-knot surface current with a 20-knot wind. The second is that the stress τ applied to the sea surface increases as the square of the wind speed:

$$\tau \cong 0.02W^2 \tag{7.7}$$

where W is the wind speed in meters per second and τ is the wind stress in dynes per square centimeter. A 10-m/s wind (nearly 20 knots) causes a wind stress of approximately 2 dynes/cm^2.

In fact, neither relationship is that simple. The "constant" varies with wind speed and surface roughness and depends critically upon how far above the sea surface the wind is measured. For winds measured at "deck height," 10–20 ft above the sea surface, Eq. (7.7) is good to within a factor of 2 and probably considerably better than that.

If one assumes a flat ocean surface with no horizontal pressure gradients and that internal friction can be ignored, the only force of any consequence is the wind stress. One can then derive a very interesting steady-state relationship from Eq. (5.26):

$$\frac{1}{\rho}\frac{\partial \tau_x}{\partial z} = -fv$$

$$\frac{1}{\rho}\frac{\partial \tau_y}{\partial z} = +fu \tag{7.8}$$

If one integrates from the surface to some depth $-Z$ beyond which the effect of wind stress is negligible,

$$\tau_x = -M_y f$$

and $\tau_y = +M_x f$ \hfill (7.9)

where

$$M_x = \int_{-Z}^{0} \rho u \, dz \tag{7.10}$$

and the units of M_x and M_y are mass per unit time per unit length. Note that a wind blowing from the north does not move water to the south, but to the *west*. Again, the Coriolis term confounds our logic. Water does not move downwind, but at right angles to the wind. For an observer facing downwind, the water moves to the right in the Northern Hemisphere; and to the left in the Southern Hemisphere.

By replacing the wind-stress term with an eddy-viscosity term, Eq. (7.8) can be written as

$$A_z \frac{\partial^2 u}{\partial z^2} = fv$$

$$A_z \frac{\partial^2 v}{\partial z^2} = -fu \tag{7.11}$$

The solution of Eq. (7.11) gives the same results as Eq. (7.9) for the total wind-driven transport, but, in addition, indicates the details of the velocity structure in the water column. The solution is a spiral in which the surface water moves at an angle of 45° to the wind (Figure 7.6).

This kind of motion is called *Ekman motion,* after V.W. Ekman, who first examined the problem in 1902. Ekman's attention was called to it by Fridtjof Nansen, who, while frozen in the Arctic aboard the

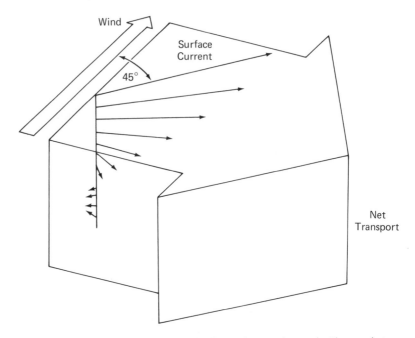

Figure 7.6. Water is set in motion by the wind. According to the Ekman relation, the effect of the Coriolis force is for each succeeding layer of water to move slightly to the right (in the Northern Hemisphere) of the one above it. The result is the Ekman spiral as shown with the net transport at 90° to the wind.

Fram, observed that when the wind blew the ice appeared to move not downwind, but at an angle of 20–40° to the right of the wind.

The variety of forces acting in the surface layer of the ocean is such that it seems unlikely that one would very often expect to observe a simple Ekman spiral, as in Figure 7.6. However, the observational evidence for a wind-driven surface current to move at some angle to the right of the wind is quite good, as is the evidence for a mass transport of water to the right of the wind, as in Eq. (7.9).

Upwelling is the term used in oceanography to describe the process by which deep water is brought to the surface. It has an importance well beond its physical significance, because the deep water carries nutrients to the surface layers where they can be assimilated by phytoplankton. Regions of upwelling are among the richest biological areas in the world.

In a formal sense, upwelling occurs wherever there is a divergent flow at the surface. Continuity requires upward vertical flow to replace the water lost by the surface divergence. The most famous

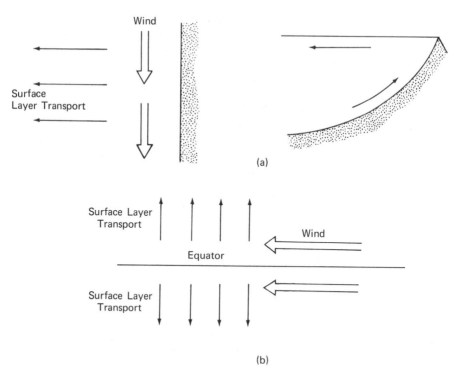

Figure 7.7. (a) Wind blowing parallel to the coast will transport the surface water offshore. This surface water will be replaced by colder water "upwelling" from below; (b) the effect of an easterly wind near the equator is to cause poleward transport of the surface water, which is replaced by colder water brought to the surface along the equator.

upwelling areas are those along certain coasts where the wind drives the surface water offshore. According to Ekman theory, the effect of wind is to drive the water to the right of the wind (in the Northern Hemisphere). Thus maximum upwelling occurs when the wind is parallel to the shore, not offshore (Figure 7.7a). Such a wind condition occurs, among other places, off the coasts of California and Peru. Colder than average surface water and higher than average biological productivity are the effects of upwelling. The region off Peru supports the highest-volume fishery in the world; this high productivity is related less to the cold rich surface waters brought from the Antarctic than it is to the wind-induced upwelling along the coast.

An examination of mean wind charts allows one to predict areas of coastal upwelling; these predictions can be verified by an examination of surface temperature charts. The cold coastal waters of Peru and California can only be explained on the basis of coastal upwelling (Figure 7.8). A striking example can be found in the Arabian Sea, where the seasonal wind pattern shifts with the monsoon (Figure 7.9). The surface temperature off Somalia drops several degrees between April and July with the advent of the Southwest Monsoon.

The effects of bottom friction, stratification, local currents, topography, and shoreline complicate the simple picture implied by Figure 7.7a. It is possible to secure general qualitative agreement between known wind systems and regions of upwelling, but the complexity of the problem continues to frustrate attempts at detailed quantitative agreement between observed winds and upwelling.

Another example of upwelling can be found along the equator, where the southeasterly trades cause an Ekman drift to the north in the Northern Hemisphere and to the south in the Southern Hemisphere (Figure 7.7b). Where the easterly component of the trades is well developed, the surface temperature along the equator in the cental Atlantic and the Pacific is often 2°C cooler than it is 100 miles on either side of the equator (Figure 7.8). On the equator, the sine of the latitude is zero, and one cannot expect Eq. (7.8) to hold, but it may apply within 100 km of the equator. There may be another explanation for the lowered surface temperature found along the equator (see the discussion of equatorial undercurrents in Chapter 8); but the evidence to date suggests that it is at least in part the result of a divergence of the surface waters resulting from a poleward Ekman transport on both sides of the equator.

Upwelling (sometimes referred to as *Ekman pumping*) can also occur in mid-ocean. The surface wind stress can result in surface divergence and upwelling in the open ocean by changes in the wind speed or direction. Horizontal divergence or convergence can be related to the wind stress vorticity:

$$\text{divergence (upwelling)} = \frac{\partial \tau_y}{\partial x} - \frac{\partial \tau_x}{\partial y} \qquad (7.12)$$

If the numerical value of Eq. (7.12) is negative, there is convergence rather than divergence and downwelling occurs. Figure 7.10 on page 128 gives examples. Note that the cyclonic winds of a hurricane result in upwelling. Colder surface water has been observed in the wake of a hurricane. The explanation may be in part related to the strong mixing of the surface waters one would expect with hurricane winds, but divergence induced by Ekman transport also might be expected to play a role.

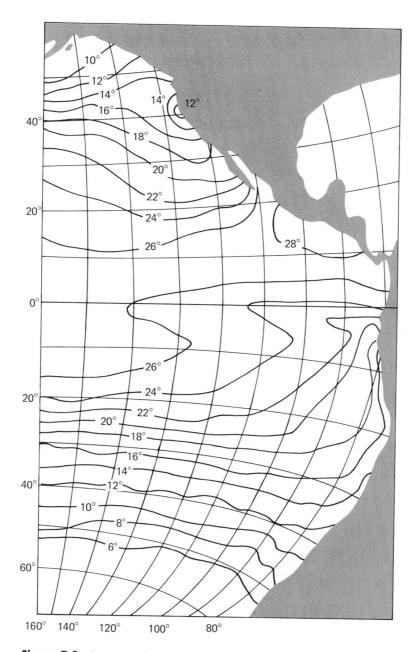

Figure 7.8. Average surface temperature for July for the Pacific. Note that the coastal water off California and Peru and Ecuador is colder than the water at the same latitude further offshore. Also note the tongue of relatively colder water along the equator. Upwelling is largely responsible for these lower than normal surface temperatures.

Inertial Motion

A final simple relation in Eq. (5.26) should be examined. Assume that somehow a particle of fluid is set in motion and that there are no horizontal forces acting on it. Pressure, wind stress, and friction are all zero. According to Newton's first law, a particle in motion will continue to move at a constant velocity in the absence of any other forces. If there are no forces, the acceleration is zero. However, a particle in motion on a rotating earth is always subject to the Coriolis force; thus the equivalent of Newton's first law is

$$\left.\begin{aligned}\frac{du}{dt} &= +fv \\[2ex] \frac{dv}{dt} &= -fu\end{aligned}\right\} \qquad (7.13)$$

Figure 7.9. (a) Strong Southwest monsoon winds of July. The solid lines with arrows indicate the average wind direction. The dotted lines indicate the average wind speed. Note that the speeds are of Beaufort 6 and 7 (22 to 33 knots) in the western Arabian Sea; (b) one effect is intense upwelling in the western Arabian Sea in July. The Southwest monsoon generally begins in May and reaches its peak in July. (c) For contrast, the surface temperature chart for April is shown.

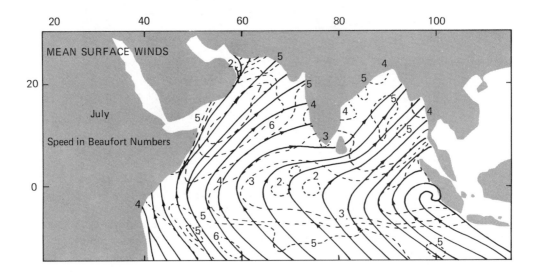

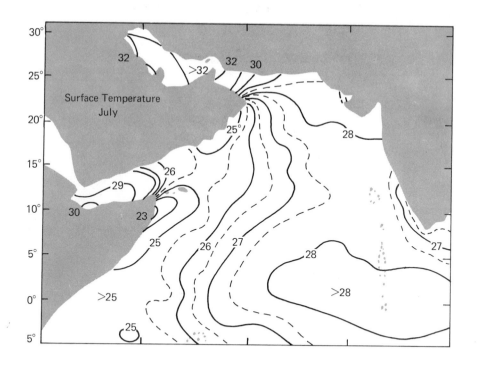

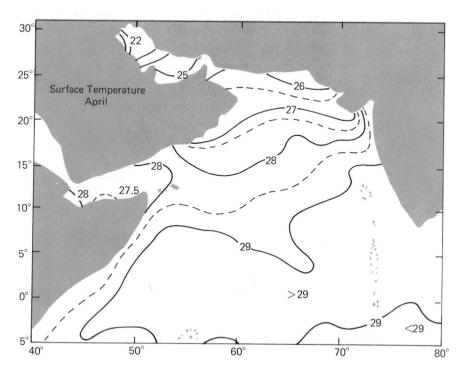

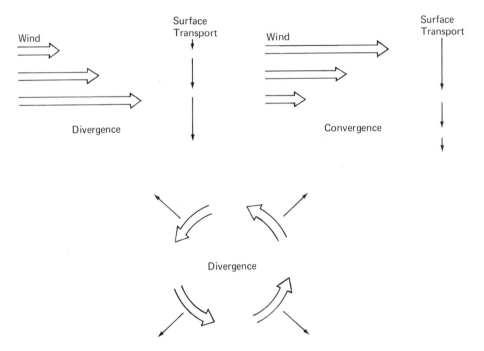

Figure 7.10. The Ekman relationship can produce a divergence of the surface currents and consequent upwelling with a cyclonic wind. The formal relationship between vorticity and divergence is given in Eq. (7.12).

It can be shown easily that this is the equation for a circle. The particle moves in a circular path with radius r at a constant speed V and with a period T_i, where

$$\left. \begin{array}{l} V = (u^2 + v^2)^{1/2} \\[2mm] r = \dfrac{V}{f} \\[2mm] T_i = \dfrac{12\ h}{\sin \phi} \end{array} \right\} \quad (7.14)$$

A particle set in motion with a speed of 50 cm/s at 42° describes a circle with a radius of 5 km and a period of 18 h. In the Northern Hemisphere the circle is clockwise; in the Southern Hemisphere, it is counterclockwise.

The motion is called an *inertial circle* Such inertial motion has been observed frequently in the ocean. One might expect to find motion that was at least partially inertial after the passage of a storm that had set the surface water in movement. However, one likely reason for the prevalence of inertial motion has nothing to do with currents at all. There is good reason to believe that internal wave energy (see Chapter 11) can be concentrated in waves with inertial periods, and that one is really observing the particle motion associated with an internal wave.

Inclined Plane with Friction

Discussing the different simple force balances in Eq. (5.26) may provide insight into the different kinds of current motion to be found in the ocean, but it does not help much in understanding why the ocean current patterns are as observed. Perhaps as much as 90% of the observed ocean currents can be described as geostrophic. However, the geostrophic equation says nothing about what drives the circulation. It merely states that the observed pressure field is balanced by the Coriolis force.

Most problems of the general circulation are beyond the scope of this text. However, perhaps some light can be shed on at least some aspects of the problem by considering the remaining two examples in this chapter.

Consider again the problem of Chapter 5 of a ball rolling down a very large inclined plane. In this example, consider the plane to be not quite frictionless, and that the friction term is in the form adopted in Eq. (5.26). The balance equation now becomes

$$\left.\begin{aligned}\frac{du}{dt} &= -gi + fv - Ju \\[2mm] \frac{dv}{dt} &= -fu - Jv\end{aligned}\right\} \tag{7.15}$$

The solution to Eq. (7.15) is shown graphically in Figure 7.11. As before, the ball is acted on by the Coriolis force. As soon as it starts to move, it is accelerated to the right. However, it is also acted upon by friction, and as a result the ball never gets back to the top of the inclined plane. It continues to move in a sinuous path, but each suc-

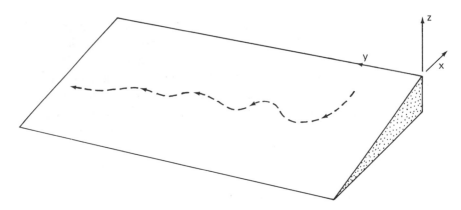

Figure 7.11. Path of a ball that rolls down an inclined plane under the influence of Coriolis force friction. Unlike the frictionless case (Fig. 5.7), each succeeding scallop is damped, until in the limiting case the ball rolls nearly parallel to the incline.

ceeding scallop is smaller than the last; eventually, it approaches a steady state where

$$
\left.
\begin{aligned}
u &= -\frac{giJ}{f^2 + J^2} \\[2ex]
v &= \ \ \frac{gif}{f^2 + J^2}
\end{aligned}
\right\}
\qquad (7.16)
$$

If friction is very small and $f \gg J$, then approximately

$$
v = \frac{gi}{f}, \qquad u = 0 \qquad\qquad (7.17)
$$

which means that the ball is in geostrophic balance and is rolling parallel to the incline.

The angle of inclination in Eq. (7.16) is

$$
\tan \theta = \frac{u}{v} = -\frac{J}{f}
$$

If $f = 10J$, then $\theta = 5.6°$, and the flow could be said to be approximately geostrophic. It is necessary that the ball continue to roll slightly downhill so that there is a loss of potential energy equal to the losses in energy due to friction. The smaller the frictional term J, the smaller the deviation from geostrophic balance.

One reason for examining this problem is to demonstrate the following point. *If* it is possible to establish and maintain an inclined plane (i.e., pressure gradient), and if the frictional forces are small compared to the pressure-gradient forces, then the resulting motion will eventually become approximately geostrophic.

Vorticity and Western Boundary Currents: The Gulf Stream

The vorticity of a particle is a measure of the spin of a particle about its axis. Vorticity is proportional to the angular momentum of a particle; clockwise spin is negative vorticity; counterclockwise spin is positive vorticity. The forces that impart spin (or vorticity) are often referred to as *torques*. It is possible to write a series of equations similar to Eq. (5.26) showing the balance of torques and rate of change in vorticity. In fact, the vorticity equations can be derived directly from Eq. (5.26) (See Appendix I). In the horizontal plane of Figure 5.1, vorticity can be formally defined as $(\partial v/\partial x) - (\partial u/\partial y)$.

For many problems that oceanographers treat, it is more convenient to think in terms of vorticity and torques than of the balance of linear forces, as in Eq. (5.26). However, it should be noted that there is nothing esoteric about vorticity balance. Any problem that can be solved in terms of vorticity can in principle be solved in terms of Eq. (5.26).

Assume a large rectangular basin with a wind blowing across it, as indicated in Figure 7.12. For purposes of comparison, this basin will represent the Atlantic Ocean and the mean wind field to which it is subject, westerlies at higher latitudes and easterly trades in the tropics. As the wind blows, the water in the basin will begin to move in a clockwise motion. The wind acts as a negative (clockwise) torque increasing negative vorticity of the water in the basin. Assume that the friction of the water in the basin can be represented by Ju and Jv, as in Eq. (5.26). This friction acts in the opposite direction from that of the water rotation. It tends to make the water spin in the opposite or counterclockwise direction. The friction is a positive torque that increases as the velocity increases. Eventually, the water will be going at such a speed that the positive torque of the friction will

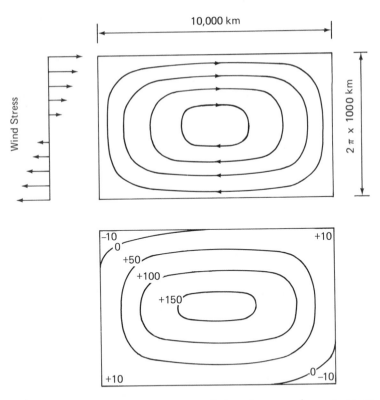

Figure 7.12. Anticyclonic wind stress applied to the sea surface sets up an anticyclonic circulation. The flow shown in the top sketch is nearly parallel to the height contours of the sea surface, which are in centimeters as shown in the bottom sketch. Under the assumption of a constant Coriolis force, the flow is symmetrical. (After Stommel, H., 1948: "The Westward Intensification of Wind-Driven Ocean Currents," *Trans. Am. Geophy. Union, 29.*)

exactly balance the negative torque of the wind stress, and the water will rotate at a constant speed even though the wind continues to blow. If the wind were to stop, there would be an imbalance, and the ocean would tend to spin the other way under the influence of the frictional torque, and eventually the rotational motion would stop.

By choosing values for the wind stress, the basin dimensions and the friction term, H. Stommel demonstrated in 1948 the circulation pattern shown in Figure 7.12. There is wind stress and friction, but the flow is nearly geostrophic. The currents flow nearly parallel to the isobars. Friction effects account for a small, downslope component to the flow, as in the case of the ball rolling on the inclined plane of the last section. The wind stress adds just sufficient energy to overcome the frictional losses and to maintain the shape of the sea surface.

The combined distribution of mass and the circulation may be compared to the energy associated with a very large flywheel, where the rotational energy of the flywheel is large compared to the rate at which energy is being added or subtracted. In the ocean a small amount of potential energy is continuously being lost by friction, but the amount lost is being replaced by the wind. The total amount of energy associated with major current systems is sufficient to maintain the circulation for many months without any input from the wind.

The most interesting aspect of the Stommel problem, however, is not the case just considered, but what occurs when one considers the Coriolis term more carefully. The first example simply represents a vorticity balance between the clockwise vorticity of the wind stress and the counterclockwise vorticity of the friction term. The earth's rotation, however, imparts an additional vorticity to a water particle moving north or south. Consider a particle moving northward at constant speed. Because the sine of the latitude increases with the latitude, the particle undergoes a clockwise spin as it moves north. Likewise, it undergoes a counterclockwise spin as it moves south (Figure 7.13). Northward-moving particles gain negative vorticity. This type of vorticity is called *planetary vorticity* since it is associated with the rotation of the earth.

By letting the Coriolis term vary with latitude, the vorticity balance is now among three terms: wind stress, friction, and the change

Figure 7.13. Planetary vorticity results from the fact that the effect of the Coriolis term increases with latitude. The effect of the Coriolis force is to subject a poleward-flowing current to anticyclonic vorticity and an equatorward-flowing current to cyclonic vorticity.

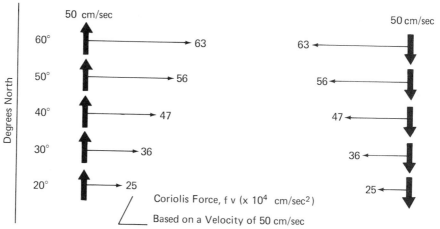

of the Coriolis term with latitude. The solution is no longer symmetric (Figure 7.14) even though the wind field is identical with that of the previous problem.

The difference between the analytical solutions described by Figures 7.12 and 7.14 can be qualitatively seen by considering the following vorticity balances. For Figure 7.12 the solution is symmetric between the clockwise wind stress vorticity and the counterclockwise frictional vorticity:

wind stress (↻) = friction (↺)

In Figure 7.14 the balance is different on either side of the ocean, since the planetary vorticity is clockwise for currents moving northward in the western ocean and counterclockwise for currents moving southward on the east shore.

Western ocean:

wind stress (↻) + planetary vorticity (↻) = friction (↺)

Eastern ocean:

wind stress (↻) = planetary vorticity (↺) + friction (↺)

The wind stress vorticity is constant on both sides of the ocean. The absolute value of the planetary vorticity and frictional vorticity terms

Figure 7.14. With the same wind stress as in Figure 7.12, but allowing for the Coriolis force to increase with latitude, the flow is no longer symmetric. The current continues to be nearly parallel to the isobars, but there is a strong western boundary current. (After Stommel, H., 1948: "The Westward Intensification of Wind-Driven Ocean Currents," *Trans. Am. Geophy. Union, 29.*)

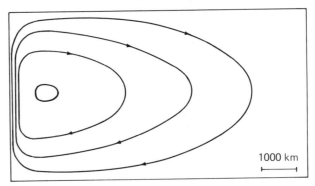

1000 km

are proportional to velocity. By inspection it is easy to show that the only solution possible is one in which the velocity in the "western ocean" is larger than that in the "eastern ocean." The ratio of the velocities depends upon the dimensions chosen for the ocean basins, frictional constants, and so on.

This westward intensification of the current can be compared to what occurs in the North Atlantic and Pacific, where the Gulf Stream and the Kuroshio are narrow, swift currents. There is nothing comparable to these narrow swift currents on the eastern side of the ocean.

Stommel made a number of simplifying assumptions in his problem, and many have since attempted more elaborate solutions of the Gulf Stream; but all solutions have one feature in common; to get a Western Boundary Current such as the Gulf Stream, it is necessary to include the change of the Coriolis term with latitude.

chapter 8

MAJOR OCEAN CURRENTS

A discussion of the complex system of water movement is facilitated by dividing the subject into more manageable pieces. Reference is often made to the surface circulation, as distinguished from the intermediate or deep circulation. Likewise, some books attempt to distinguish the wind-driven currents (mostly surface circulation) from the thermohaline currents (mostly intermediate and deep circulation). Although such tags are useful during one's introduction to the subject, they can also be misleading if taken too literally. The Gulf Stream and Antarctic Circumpolar Current are part of the wind-driven surface circulation, yet both extend to the bottom of the ocean. Although the general pattern of the surface currents of the ocean can be approximately deduced by assuming that the wind is the only driving force, few experts believe that heating and evaporation are trivial factors in determining the surface circulation patterns. The distinction between intermediate and deep circulation is an arbitrary one, and is usually based on the origin of the water masses involved. To the extent that vertical mixing is important in the ocean, such distinctions are meaningless.

In this chapter we discuss surface circulation, the major ocean currents. In the next we discuss certain features of deep and intermediate circulation, as well as some features of the interior ocean well removed from the major ocean currents. The literature in the field is vast and growing rapidly. Because the material is largely descriptive, it is not easy to summarize. No attempt is made at completeness. Hopefully, the examples chosen are representative and useful in providing insight into other aspects of the ocean circulation that are either ignored or only noted.

The gross features of the surface circulation of the ocean can be explained as wind-driven currents. There are many similarities in the circulation patterns of the Atlantic and Pacific (Figure 8.1). In both the circulation is dominated by two large anticyclonic gyres (an anticyclone rotates clockwise in the Northern Hemisphere and counterclockwise in the Southern Hemisphere). These current gyres are driven by the anticyclonic torque applied by the surface winds (see Chapter 7). The two gyres are separated by an equatorial countercurrent. In both oceans there are strong western boundary currents in the Northern Hemisphere (the Gulf Stream and the Kuroshio) and somewhat weaker ones in the Southern Hemisphere (the Brazil Current and the East Australia Current). In both the North Atlantic and North Pacific there is a cold flow from the north along the western basin: the Oyashio in the Pacific and the Labrador and Greenland currents in the North Atlantic. There is also a small cyclonic gyre, north of the main gyre in the Northern Hemisphere.

Some of the differences between oceans are due to differences in geometry; the Atlantic, Pacific, and Indian oceans are not the same shape. But some of the differences are clearly related to different wind patterns. This difference is best seen in the Indian Ocean. The circulation pattern in the southern Indian Ocean is similar in its gross features to that in the South Atlantic and South Pacific. The northern Indian Ocean circulation is clearly dominated by the monsoon. In parts of the Indian Ocean there is a complete reversal in the surface currents between the summer and winter monsoons.

As might be expected, the closer to land, the greater the deviations from the gross circulation features. There are several reasons for this. The shape of the shoreline and changes in bottom topography are probably the most important. Permanent or semipermanent ed-

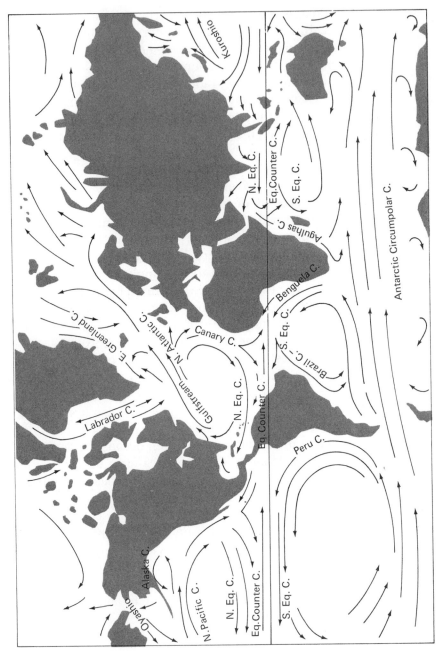

Figure 8.1. Major surface currents of the world oceans.

dies can be established as a result of the interaction of major currents with the boundaries of the ocean. Similarly, local wind patterns are more likely to deviate from the mean close to shore than in the central ocean basins. In some areas the effects of river runoff and tides are an important factor.

Finally, it should be noted that surface current charts, such as Figure 8.1, are average-value charts. No one should think he could navigate on the basis of such charts. At any given time the position of the current and its strength may be different from the average. In addition, ocean currents are turbulent. Thus the instantaneous velocity may be quite different from the mean. This is particularly true in the interior of the ocean, where the average currents are weak.

Western Boundary Currents: The Gulf Stream and Kuroshio

The part the rotation of the earth plays in establishing narrow, intense western boundary currents is discussed in Chapter 7. The western boundary currents in the Northern Hemisphere (the Gulf Stream and Kuroshio) appear to be better developed than their counterparts in the Southern Hemisphere. The reasons for this difference have yet to be satisfactorily explained. The Gulf Stream and the Kuroshio are similar in many respects. Of the two, we probably know more about the Gulf Stream.

If one considers the Gulf Stream as part of a continuous anticyclonic gyre, there is a problem in defining its beginning and end. A strong flow can be seen through the Yucatan Channel between Mexico and Cuba. This current often loops into the Gulf of Mexico before flowing through the Florida Straits. It is at this point that most people begin to refer to it as the Gulf Stream, perhaps in part because it is here that observational data become sufficient to begin to describe it in some detail. The Gulf Stream hugs the coast for some 1200 km, from Key West, Florida, to Cape Hatteras, North Carolina. Any meandering from the coast is rare. After leaving Cape Hatteras, the Gulf Stream ceases to follow a steady path. It cuts across the North Atlantic, passing below the Grand Banks; but the path varies, and sizable wavelike meanders develop in the Gulf Stream. The Gulf Stream has been tracked to 45°W some 2500 km from Cape Hatteras.

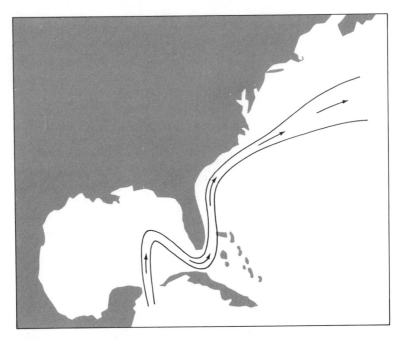

Figure 8.2. Path of Gulf Stream from Yucatan to Grand Banks. Although the Gulf Stream does not widen appreciably after it leaves Cape Hatteras, its position can range widely (see Figure 8.6).

Somewhere between the Southeast Newfoundland Rise and the Mid-Atlantic ridge the Gulf Stream ceases to be a single current (Figure 8.2).

The width of the surface Gulf Stream is between 100 and 150 km. The left edge of the stream is easily identifiable because of a horizontal temperature gradient (easier to see a few tens of meters below the surface) and a counterflow. The right edge of the Gulf Stream is less distinct. There is usually little horizontal temperature gradient, but often there is a discernible countercurrent. Surface velocities in the Gulf Stream can reach 250 cm/s, better than 5 knots. The highest velocities are to the left of the center of the current (Figure 8.3).

The Gulf Stream is in approximate geostrophic balance. To the left of the stream (facing downstream) is cold dense water. To the right is the warmer lighter water of the Sargasso Sea. Figure 8.4 shows typical temperature and σ_t cross sections of the Gulf Stream. Note that there is a slope to the constant density surfaces even below

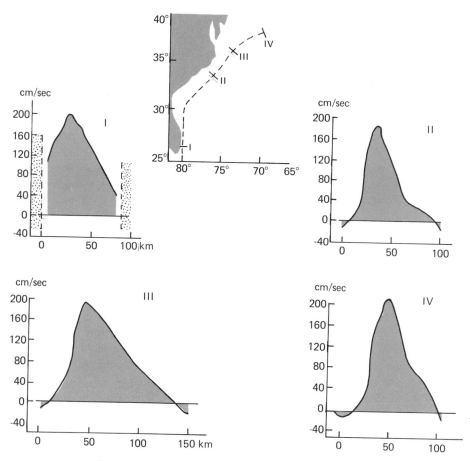

Figure 8.3. Surface velocity of the Gulf Stream. (After Knauss, J. A., 1969: "The Transport of the Gulf Stream," *Morning Review Lectures of the Sec. Int. Ocean. Cong., Moscow, 30 May to 9 June 1966*, UNESCO, Paris.)

2000 m, implying appreciable geostrophic currents at these depths. Direct observations have shown that the Gulf Stream extends to the bottom over most, if not all, of its path. Figure 8.5 is typical of the vertical distribution of geostrophic flow within the Gulf Stream.

The width of the Gulf Stream at the surface is about 100 km and increases slightly downstream. Meanders appear and grow in the Gulf Stream after it passes Cape Hatteras. Figure 8.6 is a composite picture of actual paths of the Gulf Stream observed over several months, as defined by the left edge of the stream. It is difficult to relate one

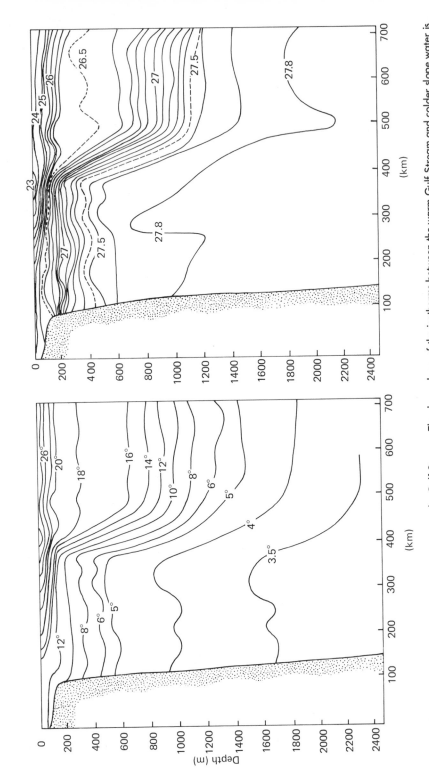

Figure 8.4. Temperature and σ_t section across the Gulf Stream. The sharp slope of the isotherms between the warm Gulf Stream and colder slope water is often referred to as the *cold wall*. The density distribution implies strong surface geostrophic flow above the sharp density gradient. (After Fuglister, F. C. and L. V. Worthington, 1951: "Some Results of a Multiple Ship Survey of the Gulf Stream," *Tellus, 3*.)

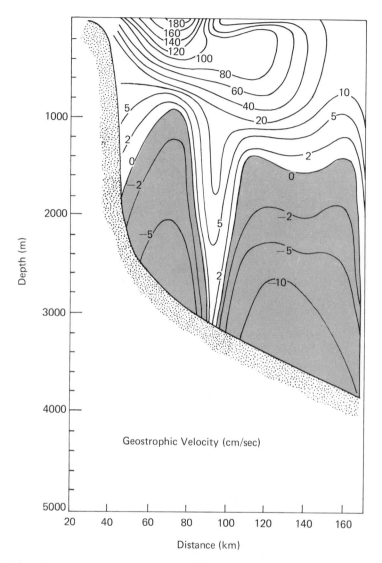

Figure 8.5. Details of the Gulf Stream flow below 1000 m are still a matter of some dispute, perhaps because they may change with time and place. This particular section shows the Gulf Stream to be narrower at depth than at the surface and the current to be displaced to the right with depth (as you look down stream). Both features are typical of most available sections. (After Richardson, P. L. and J. A. Knauss, 1971: "Gulf Stream and Western Boundary Undercurrent Observations at Cape Hatteras," *Deep-Sea Res.*, 18.)

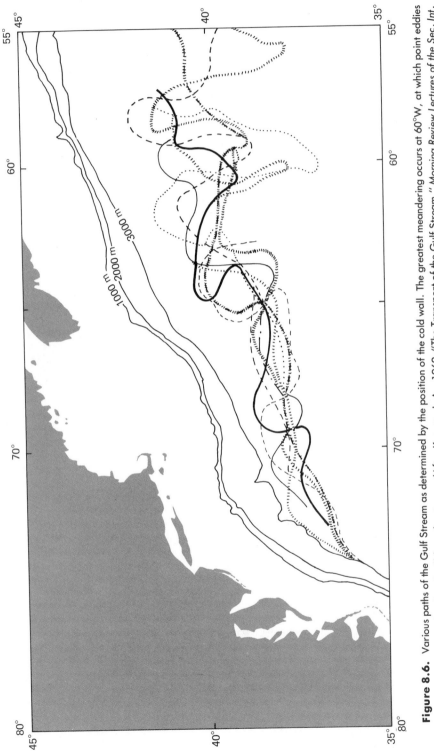

Figure 8.6. Various paths of the Gulf Stream as determined by the position of the cold wall. The greatest meandering occurs at 60°W, at which point eddies or Gulf Stream rings often break off from the stream. (After Knauss, J. A., 1969: "The Transport of the Gulf Stream," *Morning Review Lectures of the Sec. Int. Ocean. Cong., Moscow, 30 May to 9 June 1966*, UNESCO, Paris.)

path to another; however, a synoptic picture of the Gulf Stream taken every few days would show that individual meanders progress downstream.

Occasionally, these meanders grow very large, and the path of the Gulf Stream becomes difficult to follow. Often the large meanders become unstable and break off to form large cyclonic eddies south of the Gulf Stream. More rarely a meander will throw off an anticyclonic eddy north of the Gulf Stream. Figure 8.7 shows an eddy breaking off from a Gulf Stream meander over a 4-week interval. These eddies appear to last for several years. One tracked by Richardson appeared to be reabsorbed into the Gulf Stream off Florida 22 months later (Figure 8.8).

The Gulf Stream is not merely a surface phenomenon. In the Florida Straits and along the Blake Plateau, where water depths are between 500 and 1000 m, there is ample evidence of current scouring along the bottom. Current meters near the bottom have recorded speeds in excess of 1.5 knots. Evidence of the Gulf Stream extending to the bottom has been found in the deep ocean after the Gulf Stream leaves Cape Hatteras in depths greater than 4000 m.

Reliable estimates of the transport of the Gulf Stream have only recently been possible. One method is to drop a weighted float either to the bottom or to mid-depth, at which point the float releases its ballast and rises to the surface. The horizontal distance between the point of surface and point of launch, divided by the elapsed time, gives the average horizontal velocity over the depth reached by the float. A series of devices dropped across the Gulf Stream give a measure cf the volume of flow. The transport of the Gulf Stream is about 30×10^6 m³/s in the Straits of Florida. It is at least twice that at Cape Hatteras, and there is fair evidence to suggest that 1000 km downstream from Hatteras the flow of the Gulf Stream is as high as 150×10^6 m³/s (Figure 8.9).

Although the Kuroshio appears to be similar to the Gulf Stream in many aspects, it has at least one very fascinating difference. The Kuroshio can develop a larger meander that can remain stable for years. The first scientific investigation of this phenomenon was undertaken in the early 1930s. It would appear that the Kuroshio path is bimodal. It can be stable in one of two positions: with a marked meander centered at 138°E, and without the meander (Figure 8.10). The meander mode has been observed three times in the past 40 years: 1934–1944, 1953–1955, and 1959–1963.

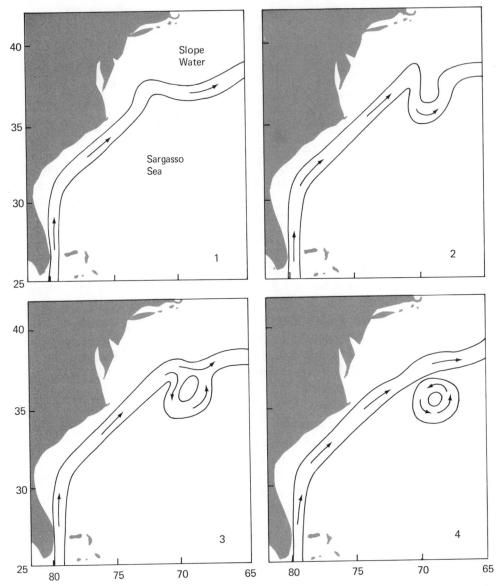

Figure 8.7. Formation of a cyclonic ring in June, 1970. Block 1 (*Upper Left*): 10–11 May, meander begins to form; block 2 (Upper Right): 20–21 May, loop forms to south; block 3 (*Lower Left*): 1 June, cold-water mass is isolated; block 4 (*Lower Right*): 8 June, ring is detached.

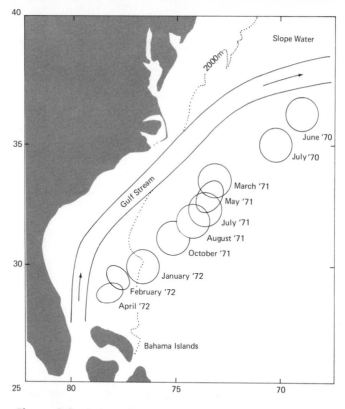

Figure 8.8. Some Gulf Stream rings have a long life. This one may have been spawned in June, 1970. It was tracked for nearly 2 years after its discovery off Cape Hatteras. (After Richardson, P. L., A. E. Strong and J. A. Knauss, 1973: "Gulf Stream Eddies: Recent Observations in the Western Sargasso Sea," *J. Phy. Oceanogr.*, 3.)

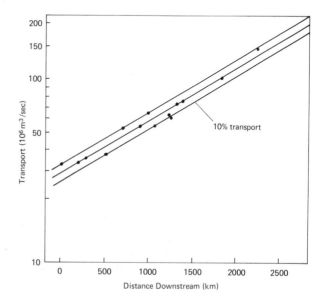

Figure 8.9. Transport of the Gulf Stream increases downstream from Miami (the 0-km point). Data on the transport beyond Cape Hatteras (the 1500-km point) are less reliable. (After Knauss, J. A., 1969: "A Note on the Transport of the Gulf Stream," *Dup-Sen Res.*, 16.)

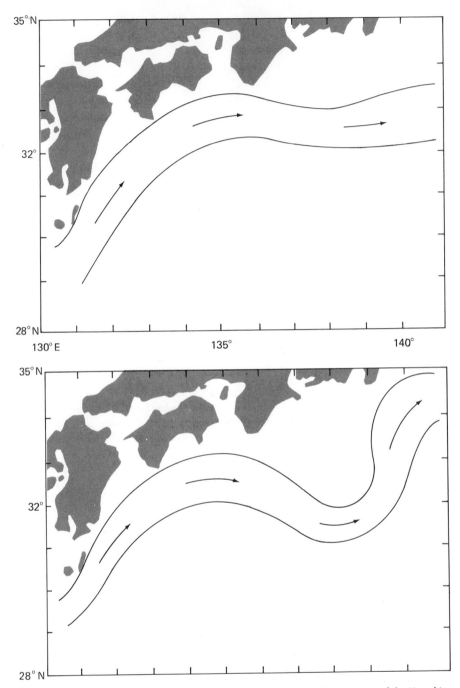

Figure 8.10. (*Above*) Two distinct and relatively stable positions of the Kuroshio; (*Facing Page*) observed tracks of the Kuroshio from March, 1956, to November 1964. The transition from one regime to the other can take several months. (After Taft, B. A., 1972: "Characteristics of the Flow of the Kuroshio South of Japan," in *Kuroshio*, eds. H. Stomel and K. Yoshida, University of Washington Press, Seattle.)

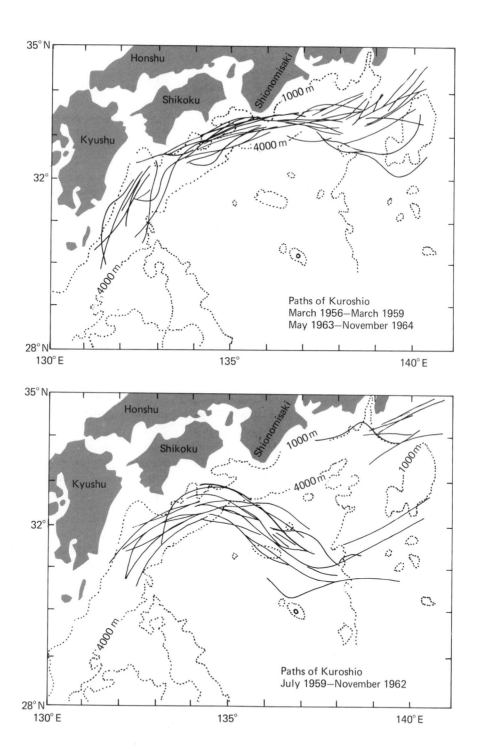

Paths of Kuroshio
March 1956–March 1959
May 1963–November 1964

Paths of Kuroshio
July 1959–November 1962

Currents Along the Eastern Sides of Oceans

The major ocean circulation may be described by large anticyclonic gyres, as shown in Figure 8.1. However, the current systems that compose a single gyre differ markedly from one part of the gyre to another. The western boundary currents, such as the Gulf Stream and Kuroshio, are narrow, swift, deep flows with comparatively well marked boundaries. The equatorward flows on the other side of the ocean basins, such as the California, Peru, or Benguela currents, are broad, weak, shallow flows whose boundaries are so poorly defined that some authors find it convenient to define different regimes as one moves offshore.

The best observed of these eastern basin currents is the California Current. The flow appears to be mostly confined to the upper 500 m. The transport is less than $15 \times 10^6 m^3/s$. A synoptic photograph of the flow at a single instant would not show a simple unidirectional flow, but rather a series of large eddies superimposed on a broad, weak equatorward movement (Figure 8.11). The speed and direction measured in the California Current at any given instance may be quite different from the average flow. A similar condition would appear to prevail in other eastern currents.

The coastal flow can be particularly complex. For descriptive purposes, the coastal flow is often separated from the broader offshore currents by a different name, and it is likely that the dynamics of the coastal current may be somewhat different from that offshore. In many eastern basins, coastal upwelling is a dominant factor in determining the distribution of properties and the local circulation. Wherever these eastern basin currents have been more than superficially studied, coastal countercurrents have been observed, sometimes at the surface, but more often beneath the surface at a depth of 1–200 m. Their persistence, the extent to which they are related to coastal upwelling, and other elements of their dynamics have yet to be determined; but it seems evident that the local topography and the extent of wind-induced upwelling can play an important role in determining the depth, position, and magnitude of a coastal subsurface countercurrent. Figure 8.12 suggests three possible regimes. Figure 8.13 is a temperature cross section in the top 300 m off Peru during a period of strong upwelling; it shows the characteristic spreading of the isotherms close to shore, which would provide a countercurrent flow similar to that in Figure 8.12c. It has been

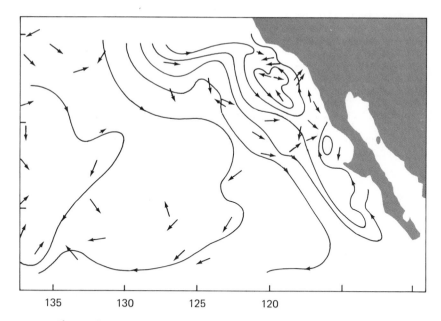

135   130   125   120

Figure 8.11. Any detailed survey of an eastern boundary current usually shows a complex flow pattern, such as this from a survey off Baja, California, in March, 1954. The lines indicate the geostrophic flow pattern as calculated from temperature and salinity observations. Independent current measurements (noted by arrows) indicated that the eddy structure was real and not an artifact of the system of observations. (After Wooster, W. S. and J. L. Reid, Jr., 1963: "Eastern Boundary Currents," in *The Sea*, ed. M. N. Hill, Interscience Publishers, New York.)

hypothesized that the two-way flow is an important factor in "conditioning" the waters off Peru to maintain the very high levels of biological productivity observed there.

Equatorial Currents

A simple way to consider the major currents of the tropics is to think of them as the counterpart of the trade-wind system. Over most of the Atlantic and Pacific there are the northeasterly trades of the Northern Hemisphere separated from the southeasterly trades, which are mostly in the Southern Hemisphere. The intertropical convergence (ITC) is an area of weak and variable winds between the two trade-wind systems, often referred to as the *doldrums* The ITC sepa-

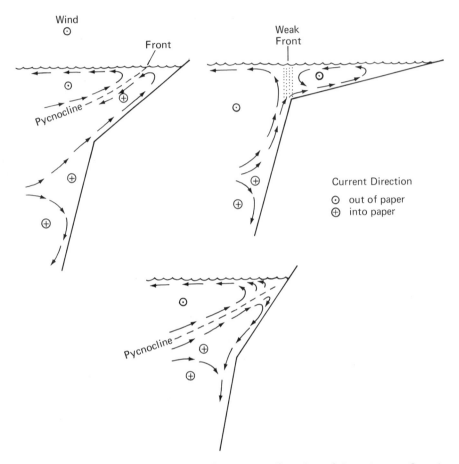

Figure 8.12. Coastal upwelling regimes (from Special Committee on Oceanic Research Proceedings, 1975: 10).

rates the wind systems of the two hemispheres and may be thought of as the climatic equator. This climatic equator is generally found between 3 and 10°N.

The major ocean currents of the tropics are a reflection of the wind system. Beneath the trades are the westward-flowing North and South Equatorial currents, which are part of the main anticyclonic current gyres of the Northern and Southern hemispheres (Figure 8.1). Between these two broad westward flows is a relatively narrow (300–500 km wide) eastward-flowing Equatorial Countercurrent (Figure 8.14). In addition, there is a very important subsurface flow at the equator, the Equatorial Undercurrent. It should also be noted that the

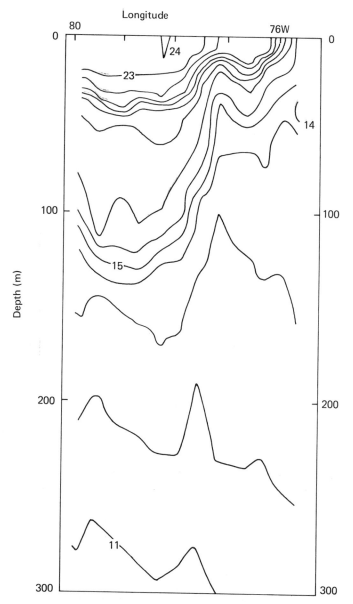

Figure 8.13. Distribution of temperature along 15°S off the coast of Peru during February, 1967. (After U.S. National Marine Fisheries Service, 1972: *EASTPAC Atlas*, Circular 330, Vol. 1, U.S. Governmenting Printing Office, Washington D.C.)

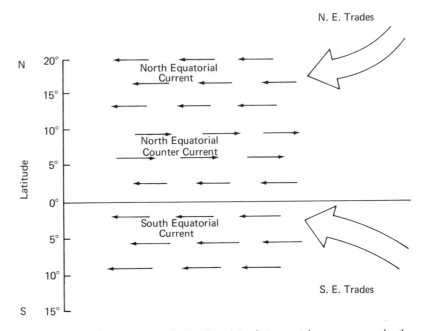

Figure 8.14. Generally, the North and South Equatorial currents are under the Northeast and Southeast trades. The Equatorial Countercurrent flows eastward in the region of the doldrums or intertropical convergence.

schematic picture of Figure 8.14 is best realized in the central Atlantic and Pacific. Both the trade winds and the equatorial current system are more complex near the edges of the ocean basins.

The tropics are characterized by a well-mixed warm surface layer and a sharp thermocline, which separates the surface layer from the cold water beneath. This thermocline also serves as a lid that separates the surface layer, which is high in dissolved oxygen and low in nutrients such as phosphate and nitrate, from the deeper waters, which are low in oxygen and relatively rich in nutrients (Figure 8.15).

Figure 8.15. A cross section of temperatures, salinity, oxygen, phosphate, and velocity at 140°W in the Pacific. In the mixed layer above the thermocline the water is high in oxygen and low in phosphate. The reverse is true below. Tongues of high salinity extend equatorward. The slope of the thermocline is consistent with a north and south equatorial current and a countercurrent between 5 and 10°N. The Cromwell Current and its effects on the thermocline can be seen on the equator. (After Knauss, J. A., 1963: "Equatorial Current Systems," in *The Sea*, ed. M. N. Hill, Interscience Publishers, New York.)

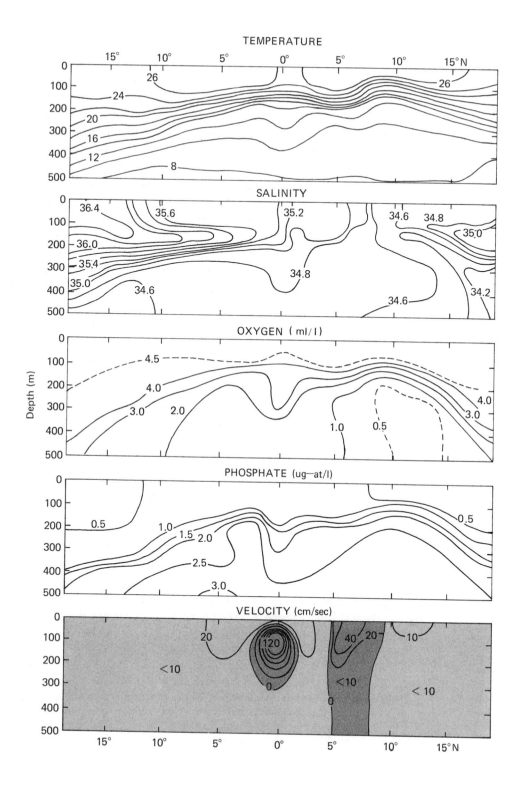

The equatorial currents are confined mostly to the surface layer, with average speeds of 25–75 cm/s. Below the thermocline, current speeds are much weaker than in the surface layer. These equatorial currents are in approximate geostrophic balance, and the slope of the thermocline reflects this geostrophic balance.

If the geostrophic flow below the thermocline is very weak or nonexistent, this requires no horizontal pressure gradient at depth. Compare Figure 7.4a with Figure 8.15. The thermocline must slope in the direction opposite to the surface. Facing downstream in the Northern Hemisphere, the thermocline slopes down to the right. In the Southern Hemisphere, it slopes down to the left. At the equator, the sine of the latitude is zero, and there is no horizontal component of the Coriolis force. However, the geostrophic approximation would appear to be valid to within half a degree of the equator.

On the equator itself is a powerful eastward-flowing subsurface current in the region of the thermocline, the Equatorial Undercurrent, or Cromwell Current, as the Pacific Equatorial Undercurrent is often called. This current is like a thin ribbon, perhaps 200 m thick and 300 km wide, with peak speeds of up to 150 cm/s (3 knots). The core of the current usually coincides with the thermocline, and it is generally centered on, or very close to, the equator. The depth of the core of the current varies from 50 m or less near its eastern terminus to 200 m or more in the western Atlantic and Pacific. On occasion the eastward flow has been observed at the surface, but usually these equatorial undercurrents are completely below the surface.

Often, but not always, there is a spreading of the thermocline with the undercurrent, as well as an apparent mixing of properties, such as oxygen, across the thermocline (Figure 8.16). It can be demonstrated that a spreading of the thermocline as in Figure 8.16 is consistent with the undercurrent being in geostrophic balance. The Cromwell Current is separated from the Equatorial Countercurrent by about 300 km (Figure 8.15). There is a sharp gradient of oxygen and phosphate values at the thermocline, except in the vicinity of the

Figure 8.16. Cromwell Current at 140°W in April, 1958. The spreading of the isotherms can be seen in the upper figure and the even more striking spreading of the oxygen isopleths in the bottom figure. Velocity contours are dashed and in intervals of 25 cm/s from 25 cm/s on the outside to 150 cm/s at the center. (After Knauss, J. A., 1959: "Measurement of the Cromwell Current," *Deep-Sea Res.,* 6.)

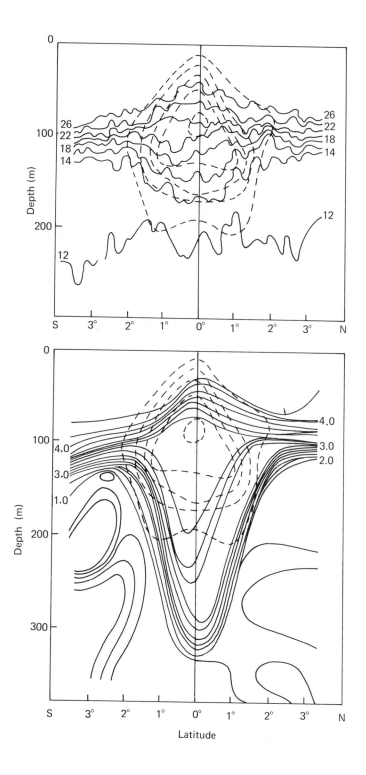

equator. A number of additional zonal currents have been suggested both north and south of the equator, but their permanence, as well as their role in the general ocean circulation, is still uncertain.

In a general way, the North and South Equatorial currents and the Equatorial Countercurrent can be shown to be wind-driven currents by an elaboration of the concepts discussed in Chapter 7. The Cromwell Current is not so easily explained. Two features are generally implicit, if not explicit, in most attempts to relate this current to the general circulation. The first is that, exactly at the equator, water does indeed flow ''downhill' and not ''around the hill'' as in geostrophic motion. Thus an east–west pressure gradient is important in maintaining the current. In both the Atlantic and the Pacific, the sea surface slopes down to the east. The slope in the Pacific averages about 5×10^{-8} (Figure 8.17).

The second idea is that geostrophy contributes to the stability of an eastward flow along the equator and to instability for a westward flow. If an eastward current is perturbed either north or south of the

Figure 8.17. Slope of the sea surface in the Pacific along the equator can be found by calculating the height of the sea surface in dynamic meters, assuming no horizontal pressure gradient at 1000 m. The highest slope corresponds with the region of highest eastward wind stresses. (After Knauss, J. A., 1963: "Equatorial Current Systems," in *The Sea*, ed. M. N. Hill, Interscience Publishers, New York.)

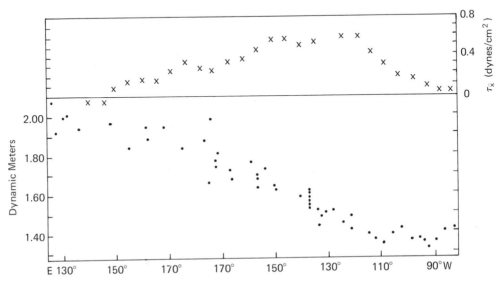

equator, the Coriolis force tends to drive it back toward the equator. However, the Coriolis force would tend to drive farther poleward a westward-flowing current (Figure 8.18). There is considerable evidence to indicate that the Equatorial Undercurrent is not always centered on the equator. On a number of occasions the core has been observed as far as 100 km from the equator. The explanation for this meandering is presumably related in some way to changes in the winds and a consequent readjustment of the density and pressure field in the vicinity of the undercurrent; but a truly satisfactory explanation is not yet available.

The wind system in the Indian Ocean is dominated by the monsoon. Equatorial winds are primarily meridional rather than zonal and reverse with seasons. They are northerly November through March and southerly May through September. There is always a South Equatorial Current, south of 5°S, stronger during the southerly monsoon than in the northerly monsoon. An Equator Countercurrent is also found at all seasons and is south of the equator. The North Equatorial Current reverses with seasons, flowing westward from November to March and eastward the rest of the time. As such, it merges with, and is indistinguishable from, the Equatorial Countercurrent (Figure 8.19).

There appears to be a slight east–west pressure gradient in March and April at the depth of the thermocline (higher pressure to the

Figure 8.18. The Coriolis force acts as a stabilizing force for an eastward jet along the equator. If the current moves away from the equator, the effect of the Coriolis force is to turn it equatorward. In the same way, the Coriolis force acts as a destabilizing effect for a westward-flowing jet. If it strays from the equator, the Coriolis force will drive it further poleward.

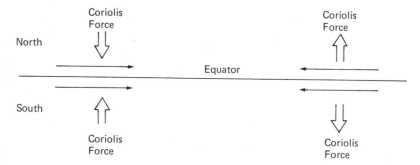

west), which gives rise to the Equatorial Undercurrent during this period. This pressure gradient is either absent or reversed in July and August, and the Equatorial Undercurrent is missing.

Antarctic Circumpolar Current

The shallow (100 m or less) Bering Straits sill between the USSR and Alaska effectively closes off flow between the Atlantic and Pacific via the Arctic. The Drake Passage between South America and the Antarctic continent is 300 miles wide and 3000 m deep, thus assuring water movement between the Atlantic and Pacific. The Antarctic Circumpolar Current that flows eastward through the Drake Passage

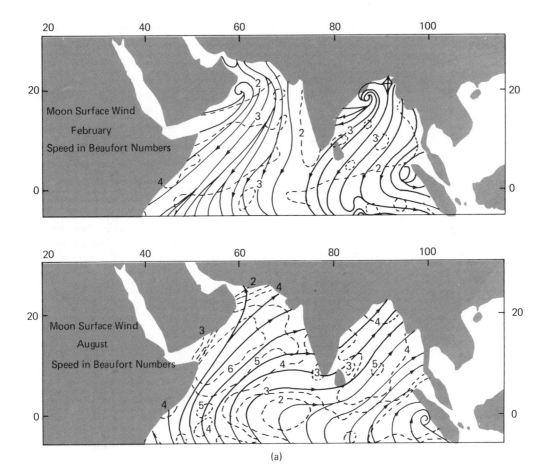

(a)

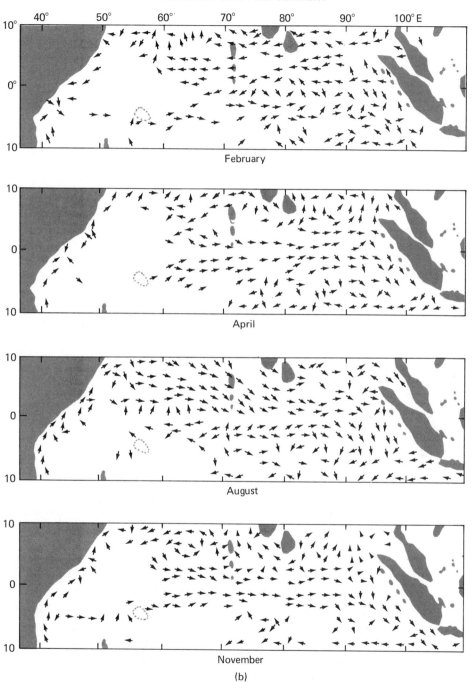

February

April

August

November

(b)

Figure 8.19. (a) Monsoon winds change about 180° between February and August (the mean wind speeds are shown by the dotted lines), (b) which results in large seasonal changes in the surface currents.

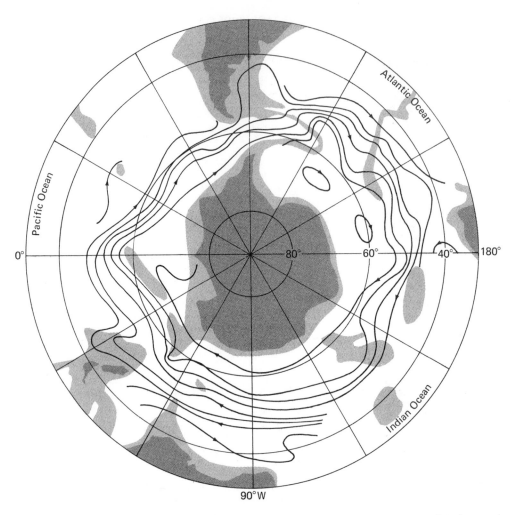

Figure 8.20. The Antarctic Circumpolar Current or West Wind Drift extends around Antarctica. Close to shore there is often a current reversal, the East Wind Drift. The lightly shaded areas represent water depths of less than 3000 m.

extends to the bottom, and has a transport of about 200×10^6 m³/s, which makes it the largest current in the world.

The Antarctic Circumpolar Current is driven by the prevailing westerly winds. Its speed and transport are thought to be a balance between the surface stress applied by the winds and frictional losses to the bottom. The apparent influence of bottom topography on its path can be demonstrated. The current tends to move southward as the water deepens and northward as it shallows. Figure 8.20 shows the axis of the Antarctic Circumpolar Current and the prevailing bottom topography.

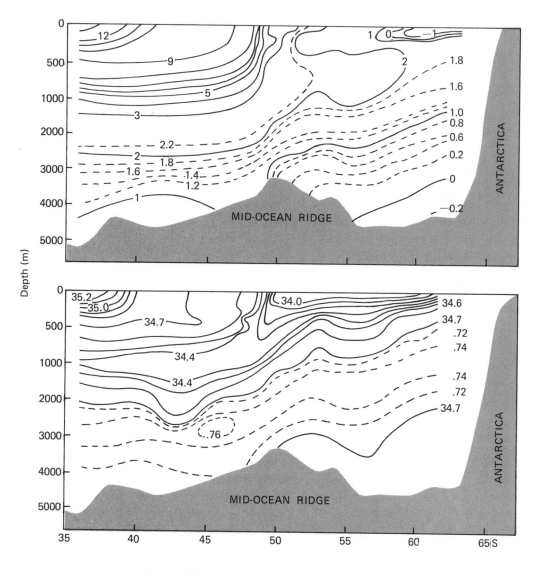

Figure 8.21. Distribution of temperature and salinity along 115°E in the Antarctic Ocean. The slope of the isotherms suggests that the Circumpolar Current extends all the way to the bottom. (After Gordon, A. L., 1971: "Introduction: Physical Oceanography of the Southeast Indian Ocean," in *Antarctic Oceanology II, The Australian-New Zealand Sector,* ed. D. E. Hayes, Antarctic Research Series, American Geophysical Union.)

Evidence of the Antarctic Circumpolar Current extending to the bottom can be seen in cross sections of temperature and salinity. The slope of the isotherms and isohalines between the Antarctic continent and 47°S in Figure 8.21 implies a geostrophic flow to the east at all

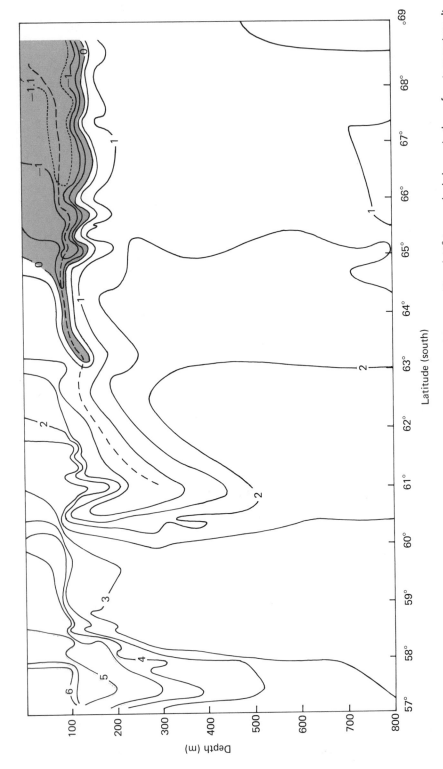

Figure 8.22. Surface manifestation of the Polar Front (or Antarctic Convergence) between 60 and 61° is a marked change in the surface temperature. It also signals the termination of a subsurface temperature minimum that extends equatorward. (After Gordon, A. L., 1971: "Antarctic Polar Front Zone," in *Antarctic Oceanology I*, ed. J. L. Reid, Antarctic Research Series, American Geophysical Union.)

depths. Current meters placed 300 m off the bottom in the Drake Passage recorded speeds of 3–9 cm/s in a generally eastward direction.

The northward edge of the Antarctic Circumpolar Current coincides approximately with the Antarctic Polar front zone, often called the Antarctic Convergence, and sometimes referred to as the Antarctic Divergence. The zone separates the Antarctic and Subantarctic surface waters. It can be recognized by a rather large gradient in surface temperature, and also as the northern terminus of a lens of water with a temperature minimum that extends equatorward from Antarctica (Figure 8.22). That this zone can be referred to as both a "convergence" and a "divergence" is some measure of the uncertainty that has characterized the various attempts to explain the dynamics of the polar front zone. The position of the polar front can shift. It has been suggested that its position is influenced by bottom topography. At times the polar front appears to extend over a width of 100 km or more. The terms "secondary" and "primary" polar fronts have been used to describe specific features within this broad band.

chapter 9

DEEP CURRENTS
AND OTHER OCEAN CIRCULATION

In this chapter we consider a wide range of ocean processes, from the formation of bottom water to the characteristics of the surface circulation in central ocean gyres far removed from major currents. The subjects have a common thread in that each helps to explain some aspects of the observed temperature and salinity structure of the ocean. As in Chapter 8, no attempt is made at completeness, but most of the key oceanic features of temperature and salinity structure are at least noted.

Thermohaline Circulation and "Core" Analysis

The circulation below the major wind-driven currents has often been referred to as *thermohaline circulation*, because the factor determining the flow patterns is the density distribution, which in turn is determined by the temperature and salinity. One of the simplest pictures of this circulation is to think of a north–south ocean basin as an onion in which the layers of the onion correspond to lines of constant σ_t (more properly σ_θ). The cold high-density water at the polar latitudes sinks the farthest; the less dense water at lower latitudes sinks to intermediate depths (Figure 9.1a). What actually occurs is much more complex (Figure 9.1b); but by detailed analysis of the temperature and salinity structure along lines of constant (or nearly constant) density, the "flow" of water in the deep ocean can often be inferred. The word "flow" is in quotation marks because usually such an analysis cannot distinguish whether the water is being advected by a steady current or is being diffused through the area by large-scale mixing processes. More critically, it is usually not possible

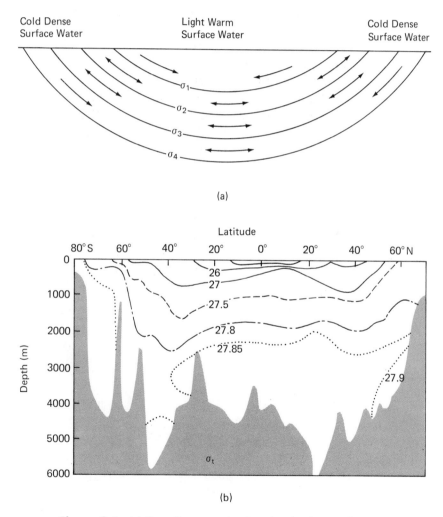

Figure 9.1. (a) To a first approximation, the distribution of temperature and salinity in the ocean can be accounted for by assuming that the water takes on its σ_t characteristics at the surface and then slides along constant σ_θ surfaces; (b) the density structure of the Atlantic bears some resemblance to this simple picture.

to say anything about the rate of flow by such an analysis. One such example is the tracing of the outflow of the Mediterranean Sea across the Atlantic Ocean. Because of its high salinity (~38.0%), Mediterranean water is denser in spite of its relatively high temperature (~13°C) than that found in any of the major ocean basins. As this warm, highly saline water flows out across the Straits of Gibraltar, it begins to sink; but as it sinks, it mixes with the colder, less saline and lighter water of the eastern North Atlantic. This mixture finds its equilibrium depth at about 1100 m, and the "core" of high-salinity water begins to spread across the North Atlantic (Figures 9.2 and 9.3).

   To know more about the spreading (e.g., does the sinking water tend to curve north or south as it leaves the Mediterranean) requires comparing a number of vertical profiles, such as Figure 9.2, radiating from the Straits of Gibraltar at different angles, or a single horizontal cross section drawn at the depth at which the water is spreading most rapidly (Figure 9.3). The high-salinity water of the Mediterranean is so clearly represented in the North Atlantic that it can be easily traced in a variety of ways. A more subtle analysis concerns the flow of the cold dense water of the Norwegian Sea through the various passages in the Denmark Strait and across the ridges running from Greenland to Scotland. The heaviest water comes through the Denmark Strait. Tracing the origin of the Antarctic Bottom Water back to the Weddell Sea is more difficult. Figure 9.4 is a plot of salinity on a surface of constant potential temperature. The flow from the Norwegian Sea is evident, as is the Antarctic Bottom Water, whose origin is

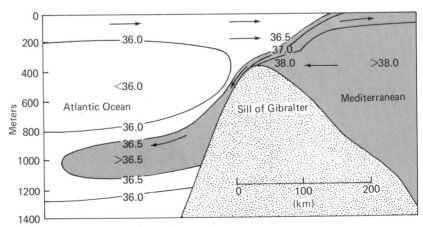

Figure 9.2. Dense, highly saline water of the Mediterranean flows into the North Atlantic over the sill at the Straits of Gibraltar. As it sinks it mixes with the surrounding water and reaches a density equilibrium at about 1000 m.

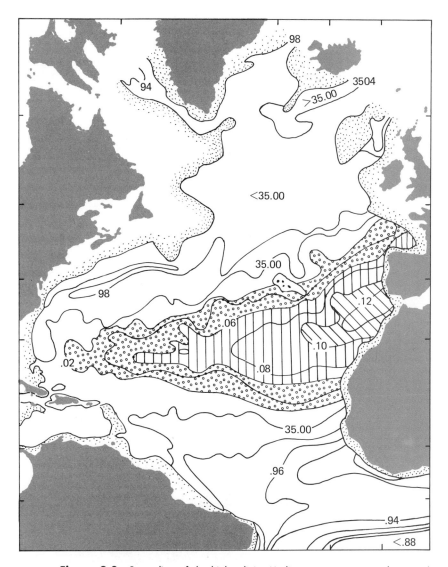

Figure 9.3. Spreading of the high-salinity Mediterranean water can be traced across the entire Atlantic. (After Worthington, L. V. and W. R. Wright, 1970: "North Atlantic Ocean Atlas," *Woods Hole Oceanographic Institution Atlas Series,* Vol. II, Woods Hole, Mass.)

in the Weddell Sea and which moves up the western basin of the South Atlantic.

As presented, the Norwegian overflow water and that from Antarctica appear almost as obvious as the Mediterranean overflow. However, the salinity anomalies involved in the latter analysis are in units of 0.01‰ of salinity, whereas for the Mediterranean an accu-

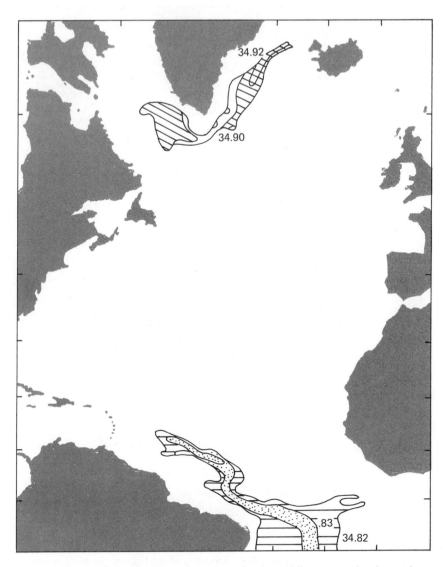

Figure 9.4. Careful analysis of small salinity differences in the deep Atlantic reveals the Antarctic Bottom Water moving up from the south and the Norwegian Sea Water moving through the Denmark Strait. (After Worthington, L. V. and W. R. Wright, 1970: "North Atlantic Ocean Atlas," *Woods Hole Oceanographic Institution Atlas Series,* Vol. II, Woods Hole, Mass.)

racy of 0.05‰ is sufficient. When trying to infer flow patterns from anomalous values of temperature, salinity, or some other conservative or semiconservative property, such as dissolved oxygen, silicon, and the like, one must be very careful. A single two-dimensional plot can be misleading, and if the method of presentation is poorly chosen, erroneous conclusions are possible. One must always be prepared to ask if the interpretation would be different if the data were plotted differently. If the answer is yes, it is important to understand the reason why.

Formation of Water Types (Masses)

The characteristic temperature and salinity distribution of such deep and intermediate water masses as the Antarctic Bottom Water, Mediterranean Water, Norwegian Sea overflow water, and others is determined by the surface condition at its place of origin and by the mixing with surrounding water once it begins to sink. Temperature and salinity are conservative properties; that is, there is no appreciable source or sink of heat or salt in the interior of the ocean. The salinity is determined at the surface as a balance between evaporation and precipitation, and the temperature is determined in the upper 100 m or less by the balance among the various terms in the heat budget equation (Chapter 3). Once the water leaves the surface layer, its temperature and salinity are fixed and are only changed by mixing with water of different temperature and salinity characteristics. The only significant exception is the flow of heat from the interior of the earth. Although relatively small, it can play a significant role in altering the temperature characteristics of the near bottom water.

One might assume the densest water would also be the coldest, but this is not the case. The densest water is found in isolated evaporation basins, such as the Mediterranean and Red Sea. However, this dense water does not fill the deep ocean basins; rather, the relatively small amount that flows out through Gibraltar mixes with surrounding water, which dilutes the salinity and reduces the density. The water finds its own density level at about 1100 m and spreads out in all directions (Figures 9.2 and 9.3). The formation of bottom water, such as the Antarctic Bottom Water and the Norwegian Overflow Water, is associated with ice formation. Water cooled to the freezing point is denser than similar water that has not yet been cooled. The freezing process increases the salinity of the adjacent unfrozen water, which in turn increases the density still further. There remain a number of perplexing problems concerning the role of freezing ice in formation of this cold, relatively high salinity water, not the least of which is the difficulty of actually observing the process. The origin of the Antarctic Bottom Water is to be found somewhere beneath the ice sheet of the Weddell Sea. Apparently, the water at "formation" has a temperature of about $-1.9°C$ and a salinity of 34.6‰. As it slides down the shelf, it mixes with the surrounding water, including the "warm deep water" with a temperature of 0.5°C and a salinity of 34.68‰, and emerges as Antarctic Bottom Water with a potential temperature of about $-0.3°C$ and a salinity of 34.66‰. This cold relatively low salinity water can be traced along the bottom up the western North Atlantic well across the equator (Figure 9.4). However, by the time it has traveled that distance it is both warmer and saltier because of mixing with the overlying water of the North Atlantic.

Apparently, no deep or bottom water is formed in the North Pacific, even in the presence of ice formation each winter in the Bering Sea. Although the Aleutian Ridge is a partial barrier for deep-water flow into the North Pacific from the Arctic, passages are available. The primary reason for the lack of deep-water formation in the North Pacific is that the surface salinities in the Bering Sea are sufficiently low that, even after freezing, the underlying water is not dense enough to displace the water already on the bottom.

The water found along the bottom in the Indian and Pacific oceans has no simple origin. It has been labeled Pacific Common Water and appears to be a mixture of various types of water found in the Atlantic and Antarctic. It enters the Indian and Pacific oceans from the south and is found at all depths below 2500 m.

Deep Western Boundary Currents

Although the origins and characteristics of the deep water are reasonably well known, much less is known about deep currents and advection. Until comparatively recently one could still question whether deep currents played an important role or not. Perhaps heat and salt are being transferred by eddy diffusion, and there is no net transfer of mass or momentum. In the absence of any measurements of deep currents, such a possibility cannot be dismissed out of hand. In many parts of the ocean, this matter is still not settled.

In recent years considerable evidence has accumulated of significant current flow in deep water along the western boundaries of the North Atlantic, the South Atlantic, and the South Pacific. All three western boundary currents flow equatorward. The evidence is of several kinds. A careful study of temperature and salinity distributions indicates a tendency for tongues of water to push down the western boundaries of ocean basins, as shown in the flow of Antarctic Bottom Water and Norwegian Overflow Water in Figure 9.4. Additional evidence can be seen in Figure 9.5, a cross section in the South Pacific off New Zealand. The slope of the isotherms in Figure 9.5

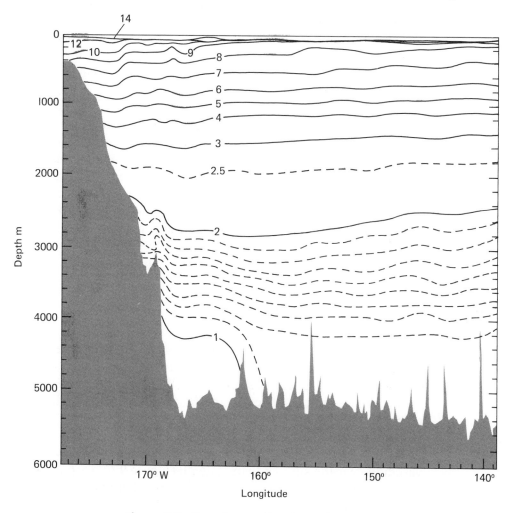

Figure 9.5. Core of water of less than 1.2°C can be seen along the western edge of the basin. This cross section is at 43°S, and geostrophic considerations indicate that the flow of the thin deep cold water should be equatorward. (After Warren, B. A., 1970: "General Circulation in the South Pacific," in *Scientific Exploration of the South Pacific*, ed. W. Wooster, National Academy of Sciences, Washington, D. C.)

implies a horizontal pressure gradient. Geostrophic calculations made on the basis of data from this section suggest a northward flow close to the bottom of several centimeters per second. The calculated transport is about 15×10^6 m³/s.

Direct evidence has been accumulated by recording current meters suspended close to the bottom. Zimmerman has reported flows of about 10 cm/s averaged over 1 week to 10 days for the western boundary current that flows down along the East Coast of the United States. These deep western boundary flows are capable of moving sediment, and indirect evidence of their presence has been found along the western slopes of both the North and South Atlantic on the basis of sediment analysis. In the North Atlantic, sediment that originates in the Gulf of St. Lawrence has been found along the continental slope as far south as South Carolina.

It has been suggested that most north–south advection is by these deep western boundary currents, and that most east–west movement is by turbulent diffusion. It remains to be seen whether such a hypothesis is substantiated. One problem with the development of any theory of the deep circulation is that we have little evidence of where the water returns to the surface. The sources of deep water are few and are known. However, the water that sinks from the surface must be replaced. It is not known whether this return is at a few spots, such as regions of upwelling, or whether the water returns slowly over such a wide area that it cannot be detected.

Characterization of Water Masses and Types

A number of attempts have been made to characterize water masses; one of the oldest is the temperature–salinity diagram (or T–S diagram). In any given area a plot of all temperature versus salinity data, from near the surface to the bottom, has very distinctive and reproducible characteristics (excluding the top 100 m where seasonal temperature changes may occur). All data from a given region of the ocean can be expected to fall within a certain envelope, with greater variation near the surface than at depth (Figure 9.6). Inspection of the shape of the curve allows one to distinguish between water masses. Figure 9.7 shows the T–S envelopes in which all data from different regions of the oceans can be expected to fall.

The temperature–salinity characteristics of the deep water in the Atlantic are considerably different from those of the Pacific and Indian oceans. In the Pacific and, to a slightly lesser extent, in the

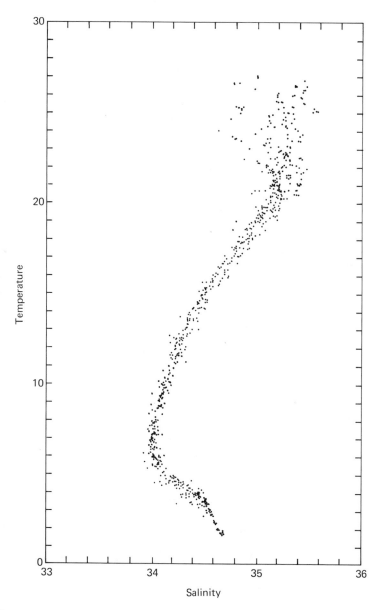

Figure 9.6. For a given region, all temperature and salinity data will fall within a given envelope of a T–S diagram. Data that do not are more likely to indicate problems with instrumentation than new discoveries about the ocean. (Courtesy of W. J. Emery.)

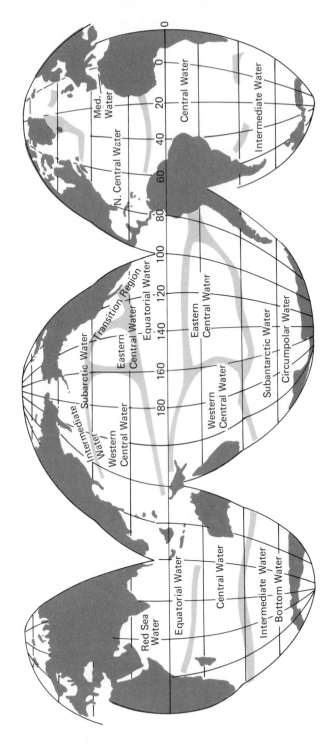

Figure 9.7. Temperature–salinity diagrams for the world oceans. The world-ocean chart indicate regions where central water masses are found. (After Sverdrup, H. U., M. W. Johnson, and R. H. Fleming, 1942: *The Oceans,* Prentice-Hall, Inc., Englewood Cliffs, N.J. and U.S. Navy Hydrographic Office, 1956: Special Publication 11.)

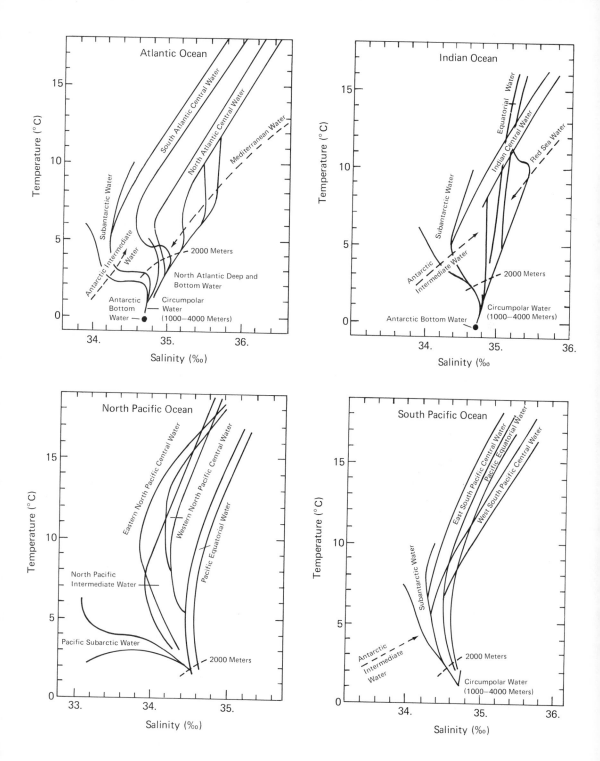

Indian Ocean the deep water is remarkably uniform. This is because all of the water in the deep Pacific has a common origin, as has nearly all the water in the Indian Ocean. There is a small outflow of Red Sea water in the Indian Ocean analogous to the outfall from the Mediterranean into the Atlantic. Nearly all the deep water in the Pacific and Indian oceans gains its temperature and salinity characteristics in the great mixing basin of the Antarctic Circumpolar Current.

The Atlantic, on the other hand, has four sources of deep water. These are the previously noted Antarctic Bottom Water, the Mediterranean water, and the Norwegian Sea water. Deep water is also formed at least some years in the Labrador Sea. The T–S characteristics of these four sources are slightly different. Furthermore, one can detect differences between the Norwegian Sea water that flows through the Denmark Straits and that which comes across the Iceland–Scotland ridge. As a result, the differences among the T–S diagrams for the deep water of the Indian and North and South Pacific (stations E through I) are relatively small compared to that found between the North and South Atlantic (stations A through D) in Figure 9.8.

Another way of characterizing ocean waters is by constructing bivariate histograms, as has been done by Montgomery and his colleagues. These show the volume of water of given temperature and salinity. It is evident that the deep Atlantic is less well mixed than the Pacific. Although the Pacific has twice the volume of the Atlantic, it takes only five classes to describe 50% of the water in the Pacific, whereas it takes 11 classes to describe 50% of the water of the Atlantic (Figure 9.9).

One effect of the Mediterranean outfall is to make the median temperature of the Atlantic somewhat warmer than the Pacific. This difference was first noted by Buchan after analyzing the temperature data of the *Challenger* expedition, 1872–1876, the first systematic set of deep temperature observations ever collected in the world oceans. Modern temperature charts show considerably more detail but do not affect the gross difference (e.g., see Table 9.1).

Basins and Sills:
The Role of Bottom Topography
in Determining Temperature and Salinity Distributions

Bottom topography often plays an important role in determining the temperature and salinity characteristics of the deep ocean. For

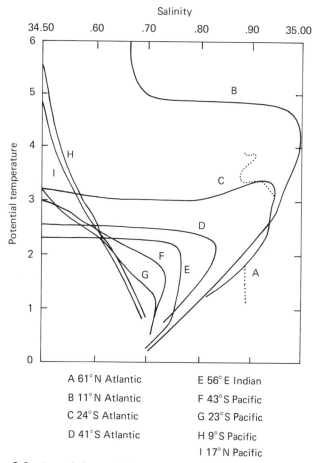

Figure 9.8. Expanded potential temperature–salinity diagram clearly shows the relatively large differences in the deep water of the Atlantic versus that in the Pacific. (In part from Reid, J. L. and R. L. Lynn, 1971: "On the influence of the Norwegian-Greenland and Wedell Seas upon the Bottom Waters of the Indian and Pacific Oceans," *Deep-Sea Res.,* 18.)

example, the Drake Passage effectively blocks most, if not all, of the Weddell Sea bottom water from flowing into the Pacific. Similarly, there are no passages by which the deep water of the Arctic Ocean can find its way into the Atlantic or Pacific. Most dramatically, the water characteristics in such isolated basins as the Red Sea and the Mediterranean are considerably different than those of the open ocean.

Wüst first suggested in 1936 that the 2°C potential temperature difference between the eastern and western Atlantic bottom water at 20°S is caused by bottom topography. The bottom water in the east-

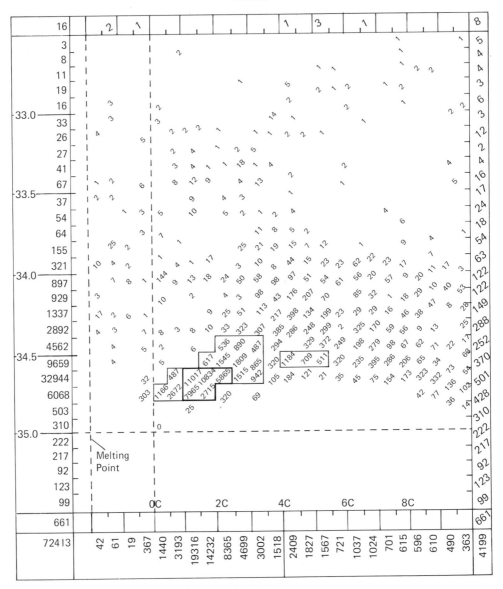

Figure 9.9. Bivariate histograms indicate the volume/of water in 10^4 km^3 found in each bivariate class 0.5°C × 0.1‰. Water outside the range of the diagram is represented by oblique numbers. Sums at bottom give the distribution by potential temperature and along the vertical by salinity. Heavy boundary encloses 50% of total, fine boundary encloses 75%. These histograms show rather dramatically the

Atlantic

relative inhomogeneity of the Atlantic compared to the Pacific which has twice the volume. (After Cocrane, J. D., 1958: "The Frequency Distribution of Water Characteristics in the Pacific Ocean," *Deep-Sea Res.*, 5. and Montgomery, R. B., 1958: "Water Characteristics of Atlantic Ocean and of World Ocean," *Deep-Sea Res.*, 5.)

Table 9.1 Distribution of Temperature, Salinity, and Specific Volume*

Ocean	Mean	Lower quartile		Median	Upper quartile	
		5%	25%	50%	75%	95%
Potential temperature (°C)						
Pacific	3.36	0.8	1.3	1.9	3.4	11.1
Indian	3.72	−0.2	1.0	1.9	4.4	12.7
Atlantic	3.73	−0.6	1.7	2.6	3.9	13.7
World	3.52	0.0	1.3	2.1	3.8	12.6
Salinity (per mile)						
Pacific	34.62	34.27	34.57	34.65	34.70	34.79
Indian	34.76	34.44	34.66	34.73	34.79	35.19
Atlantic	34.90	34.41	34.71	34.90	34.97	35.73
World	34.72	34.33	34.61	34.69	34.79	35.10
Potential specific-volume anomaly (centiliters/ton)						
Pacific	62	22	31	39	66	162
Indian	56	21	25	31	63	145
Atlantic	45	8	22	28	46	137
World	56	20	26	36	62	149

*After Montgomery, R.B., 1958: *Water Characteristics of Atlantic Ocean and of World Oceans*, Deep-Sea Res., 5.

ern Atlantic is constrained by the Walvis Ridge as well as the Mid-Atlantic Ridge. The small amount of water of less than 1.5°C that finds its way to the eastern Atlantic north of the equator arrives there by flowing up the western Atlantic and then through the Romanche trench to the eastern Atlantic (Figure 9.10).

Sturges has been able to demonstrate on the basis of small, but significant differences in the T–S diagrams from the deep basins of the Caribbean that all the recent flow into the four Caribbean basins has come from water flowing through a single channel, the Windward

Figure 9.10. (a) Bottom temperature for depths greater than 4000 m; (b) temperature cross section for the Eastern Atlantic. The effect of bottom topography is clearly seen. The blocking effect of the Walvis Ridge at 20°s is evident in the appearance of temperatures of less than 1.5°C north of the equator, which can only be accounted for by a flow into the eastern North Atlantic from the western basin. (After Wust, G., 1936: "Das Bodenwasser und Die Stratosphare in Schicting und Zirkulation des Atlantischen Oceans," *Deutsche Atlantische Expedition "Meteor" 1925–1927*, B and VI-Erster Teil, De Gruyter & Co., Berlin and Leipzig.)

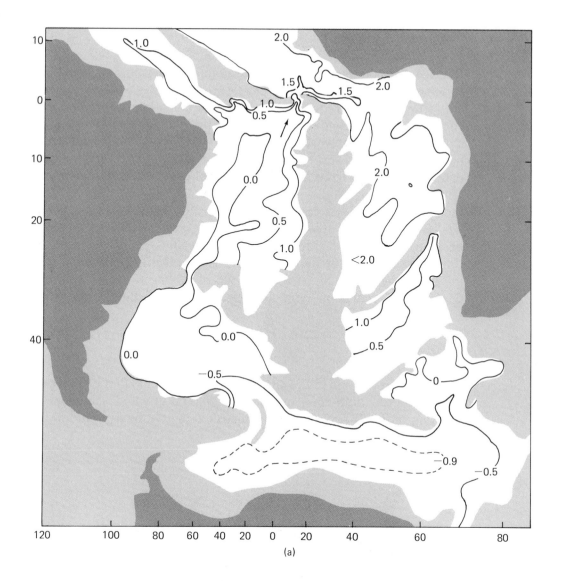

(a)

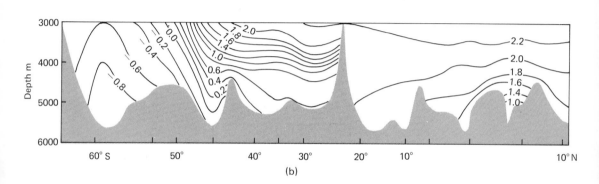

(b)

Passage. Wyrtki used a similar technique to determine the flow in a series of basins east of Australia.

The deep-sea trenches that ring the Pacific can be treated as a special case. The Mindanao trench near the Philippines has a depth in excess of 10,000 m, which is about 6000 m below the adjacent ocean basin. These deep-sea trenches are long and narrow. Temperature data are available from all deep-sea trenches; in each, the minimum temperature is at or close to the sill depth. Below the sill the *in situ* temperature increases (Figure 9.11 and Table 9.2). As nearly as can be determined, the temperature increase is adiabatic. In all these deep-sea trenches the dissolved oxygen values are comparable to those found in the surrounding water. This indicates reasonable mixing in the basin, since, in the absence of mixing, oxygen would be depleted by organisms living and decaying within the trench.

Microstructure

The introduction in the late 1960s of instruments that could accurately record continuous profiles of temperature and salinity as a function of depth revealed an unsuspected degree of structure in the

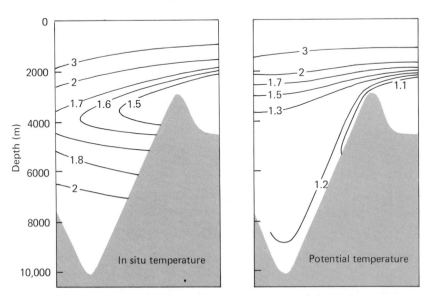

Figure 9.11. In the deep trenches that ring the Pacific, the in situ temperature increases below the sill depth; the potential temperature is essentially isothermal.

Table 9.2 Comparison of in situ and Potential Temperature in Trenches

Depth (m)	Salinity (⁰/oo)	Temperature		Density	
		In situ (°C)	Potential (°C)	σ_t	σ_θ
1,455	34.58	3.20	3.09	27.55	27.56
2,470	34.64	1.82	1.65	27.72	27.73
3,470	34.67	1.59	1.31	27.76	27.78
4,450	34.67	1.65	1.25	27.76	27.78
6,450	34.67	1.93	1.25	27.74	27.79
8,450	34.69	2.23	1.22	27.72	27.79
10,035	34.67	2.48	1.16	27.69	27.79

*After Pickard, G.L., 1975: *Descriptive Physical Oceanography*, Pergamon Press, New York.

ocean (Figure 9.12). Often the temperature does not change continuously with depth, but rather in a series of nearly discontinuous steps. At times these steps can be of the order of a few tenths of a degree and quite regular. The temperature structure is accompanied by a comparable structure in salinity, which is compensatory in the sense that to a first approximation the density versus depth curve is continuous (Figure 9.13). Some level of such microstructure seems to be nearly ubiquitous in the oceans, but the large fairly regular structure of Figure 9.13 is much more rare. The explanation for the structure is not settled, and it is possible that there is no single explanation. Certainly, some of the "horizontal layering" implied by these discontinuities can be explained on the basis that horizontal mixing is several orders of magnitude larger than vertical mixing, with the consequence that, as different types of water mix horizontally, sharp vertical gradients can be expected to occur occasionally. It has been suggested that such gradients might be a consequence of breaking internal waves. Such waves have been observed in the ocean (Figure 9.14), but whether or not they can account for the various observed microstructures is less clear.

One of the most intriguing suggestions is that the microstructure results from the fact that the rate of molecular diffusion of heat is two orders of magnitude greater than the molecular diffusion of salt (Chapter 4). For purposes of demonstration, assume a simple two-layer ocean in which the temperature and salinity gradients exactly compensate so that there is no density gradient (Figure 9.15a). Since the diffusion of heat will occur faster than the diffusion of salt, at some time in the future the gradient will look like Figure 9.15b. The

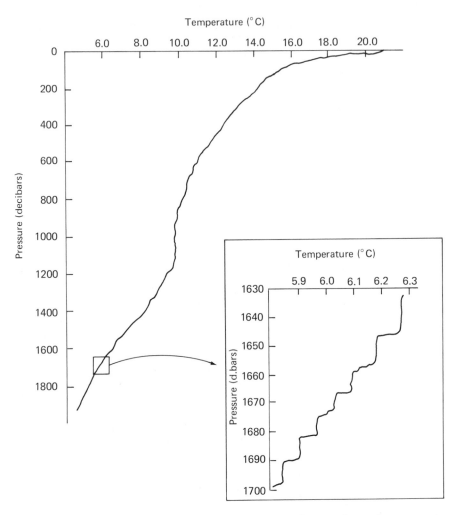

Figure 9.12. Sharp stepwise temperature gradient can be easily seen on the blown-up section of the temperatures traced from the Mediterranean Outflow Region (34°N, 11°W). The temperature steps are generally less than 0.1°C and occur at about 10-m intervals. (After Magnell, B., 1976: "Salt Fingers in the Mediterranean Outflow Region (34°N, 11°W) using a Towed Sensor," *J. Phy. Oceanogr.*, 6.)

interface will become unstable, and the saltier water will sink until it reaches a new equilibrium point. The process can be demonstrated in the laboratory, and rather than the entire layer sinking uniformly, small columns or *salt fingers* sink. Salt fingers can occur even if the interface is slightly stable initially. Scale analysis for the ocean suggests that the fingers should have a length of 20–30 cm and a

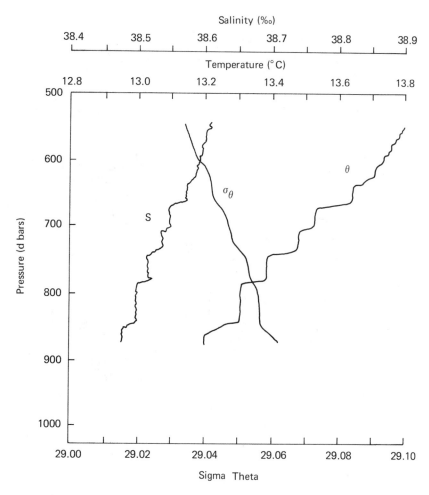

Figure 9.13. Temperature steps are usually concurrent with steps in the salinity gradient. The two tend to offset one another so that the density gradient is relatively smooth. (After Molcard, R. and A. J. Williams, 1975: "Deep Stepped Structure in the Tyrrhenian Sea," *Me. Soc. Roy. des Sci. de Leige, 7.*)

temperature difference of 0.1°C or less, with a spacing between the salt fingers of about 1 cm. Detecting the salt fingers in the ocean has been difficult, but it would appear that they have been observed in the ocean on at least a few occasions and that at least some of the microstructure in the ocean is caused by salt fingers. It is less clear, however, whether "double diffusion" plays a major role in determining the observed microstructure and the vertical mixing of the ocean.

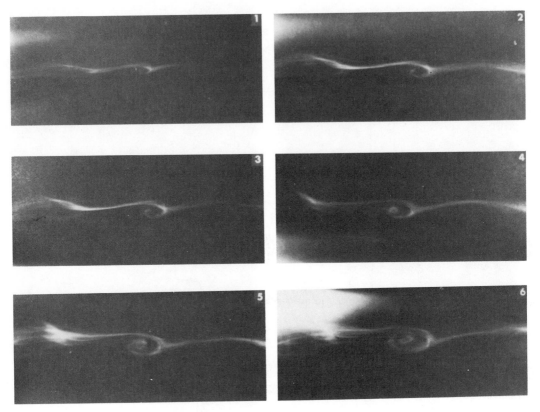

Figure 9.14. Buildup and breaking of an internal wave can be readily seen on the sequence of six photographs taken over a period of approximately 2 min at a depth of approximately 50 m in the Mediterranean Sea around Malta. The height of the wave is 30 cm. (From Woods, J. D., 1968: "Wave-induces Shear Instability in the Summer Thermocline," *J. Fluid Mech.*, Vol. 32, part 4, with permission of author.)

Mesoscale Turbulence: The Role of Eddies

The role of turbulent eddy diffusion has long presented a frustrating problem to oceanography. It has been evoked as a major frictional dissipative force in the equation of motion and in the conservation equation as a way of transferring material. Some have suggested that eddy diffusion may be more important than the major currents for the lateral transport of heat and salt. However, until relatively recently there has been little quantitative information concerning eddies in the interior of the ocean, other than the rather special case of those

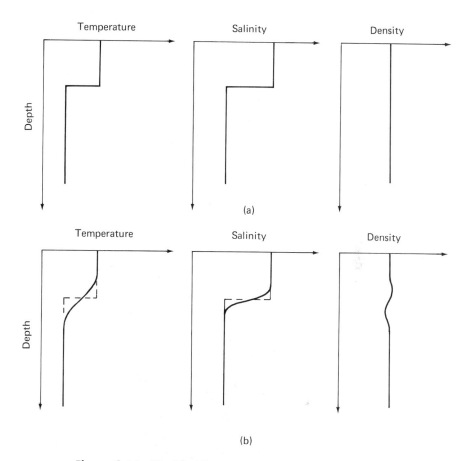

Figure 9.15. "Double diffusion" can lead to an instability. In the example, the initial density in the upper, warmer, and saltier layer is identical with the density in the lower, colder and less saline layer (a). Heat diffuses faster than salt, with the result that density gradient becomes unstable.

thrown off by the Gulf Stream (Figures 8.7 and 8.8). Recent work has demonstrated that eddy motion in the western North Atlantic does exist. An eddy with a horizontal scale of about 200 km was observed to move through a heavily instrumented area. The structure of the eddy changed during the observation period (Figure 9.16). Some of the most interesting sets of observations are those of neutrally buoyant floats that play the role of subsurface drift bottles. As can be seen in Figure 9.17, these floats appear to move independently of one another and almost at random. Although the role of eddies in ocean

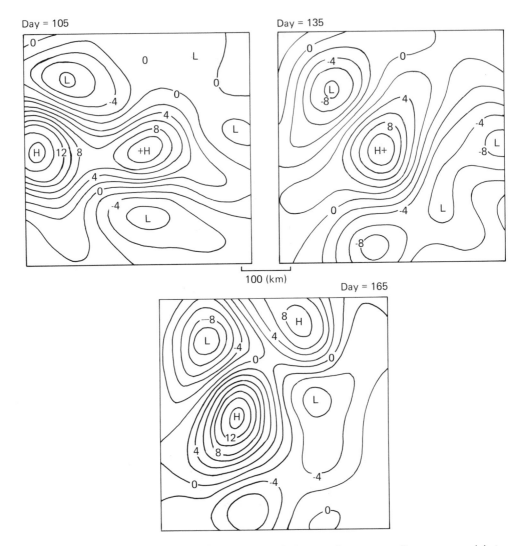

Figure 9.16. Slow change in the horizontal pressure gradient as measured during the MODE program over a 60-day period. The contours are dynamic centimeters and the depth is 150 m. The position is in the Western North Atlantic centered at 69°40′W and 28°N. Although the area was dominated by a high-pressure cell (a clockwise eddy) during the observational period, the details of the pattern changed significantly during the 60 days. (After McWilliams, J. C., 1976: "Maps from the Mid-Ocean Dynamics Experiment, I, Geostrophic Stream Function," *J. Phy. Oceanogr.*, 6.)

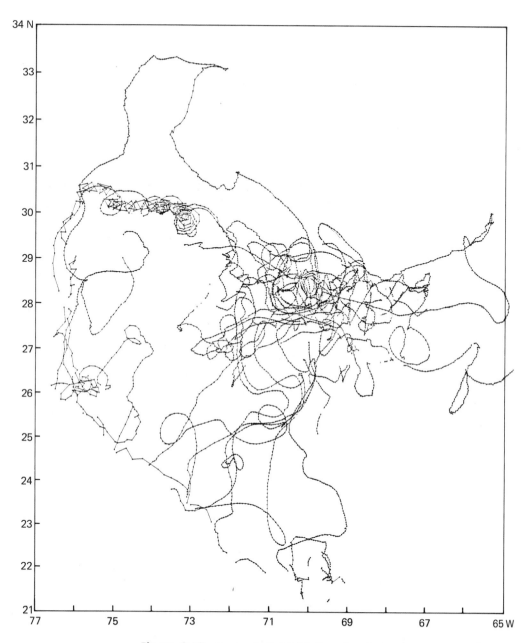

Figure 9.17. Tracks of 48 SOFAR floats launched in the vicinity of 28°N and 69°W in November, 1972, and tracked through December, 1974. The floats were at 1500-m depth. Each dot represents a daily position so that wide spacing between dots means faster movement than when the dots fall close together. Although some floats moved "together" for short periods, the overall effect is one of almost random movement. (After Rossby, T., A. D. Voorhis and D. Webb, 1975: "A Quasi-Lagrangian Study of Mid-Ocean Variability using Long Range SOFAR Floats," *J. Mar. Res.,* 33.)

circulation is still to be demonstrated, the observations themselves have opened up an entirely new scale of observational phenomena that requires description and explanation.

Langmuir Circulation

Although a large amount of theoretical work has been based on Ekman circulation and Ekman spirals, the observational evidence for Ekman spirals, at least, is minimal. Paradoxically, there is a fair amount of evidence for a form of near-surface circulation called Langmuir circulation, after the man who first described it in 1938; but there is little theoretical understanding of the process. The surface manifestations of Langmuir circulation are lines of convergence, windrows lined up nearly parallel to the wind. There is some suggestion that in the ocean, at least, the windrows are at an angle of perhaps 15° to the right of the wind (in the Northern Hemisphere where such observations have been made).

The circulation pattern appears to be a series of vortices with individual particle motion being helical as it moves downwind (Figure 9.18). There are also observations to suggest that the vortices are

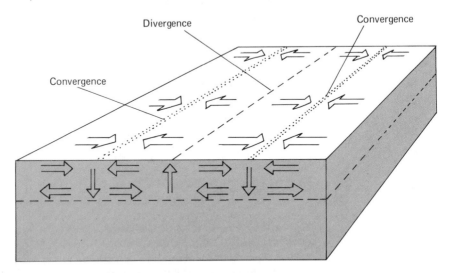

Figure 9.18. Convergence, followed by downwelling, divergence, then upwelling in a Langmuir cell is combined with movement downwind (along the line of the convergence). The result is a helical pattern for a given water particle.

asymetrical, with the clockwise vortices (as one faces downwind) being larger than the counterclockwise vortices. Such circulation has been observed and studied in lakes as well as in the ocean. Many of the ocean observations (including Langmuir's original observations) have been made in the Sargasso Sea, where the floating *Sargassum* moves toward the convergence zone and provides dramatic visual evidence of Langmuir circulation. A number of observers have commented on the rapid response of the windrows to a change in wind direction, within 10 min on at least one occasion after a 45° shift in the wind direction.

The depth and spacing of the vortices is a matter of dispute. The width of the cells is generally several times greater than the depth, and the depth may be controlled, at least in part, by the depth of the mixed layer and the stability of the water column. Apparently, the width is a function of wind speed if the mixed layer is deep enough. For example, it has been suggested on the basis of observation in the ocean that $B = 5W$, where B is the distance between windrows and W is the wind speed in length units per second. However, the depth of the vortice is apparently limited by the depth of the mixed layer; thus, in the presence of a shallow mixed layer, as in the case of lakes, the width never approaches the value of $B = 5W$.

A number of observations have been made of the downwelling velocity and a few in the region of upwelling. Downwelling velocities appear larger and are roughly 1% of the wind speed; that is, a 6-m/s wind produces a 6-cm/s vertical velocity.

chapter 10

SURFACE WAVES

Everyone has seen surface waves on water, the slight ruffling by the wind of the surface of a lake, the breaking of the surf on a beach. The slow oscillation of sea level in a harbor or lake after the passage of a line squall is also a wave, as is the tide. Waves are periodic; a wave's period and its height are perhaps its simplest characteristics. Cat's paws, the waves caused by a gust of wind on a calm lake, have characteristic heights of 1 cm and periods of 1 s or less. The surf breaking on a beach is measured in meters with periods of 4–12 s. Harbor oscillations have periods measured in minutes, and tides are measured in hours.

Attempts have been made to summarize the average characteristics of the ocean surface waves. Figure 10.1 is one such attempt. The horizontal scale is wave frequency ω, where $\omega = 2\pi/T$, and T is the wave period. The vertical scale is the square of the wave amplitude. The amplitude squared is a measure of the amount of energy in a wave. As can be seen in Figure 10.1, most of the energy is in the range of 4–12 s, a fact apparent to every surfboard rider. The only other

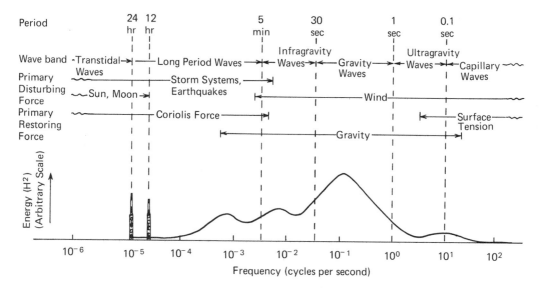

Figure 10.1. Estimate of the amount of energy in surface waves. The 12- and 24-h tides are sharply defined. Most energy is in the 4- 12-s gravity waves. (After Kinsman, B., 1965: *Wind Waves*, Prentice-Hall, Inc., Englewood Cliffs, N.J.)

periods with appreciable energy are the semidiurnal and diurnal tides. However, it is also of interest to note that there is at least some wave energy in all periods, from a second to more than a day.

Wave Characteristics

Waves can be characterized by their period, length, speed, and amplitude or height. In a simple wave, as in Figure 10.2, the length is the distance between two crests. The wave height is the vertical distance from trough to crest and is twice the amplitude. The period is the time between passage of two successive crests past the same point. Thus the speed of the wave is

$$C = \frac{\Lambda}{T} \tag{10.1}$$

It is often convenient to refer to wavelength in terms of a wave number κ, and period in terms of wave frequency ω.

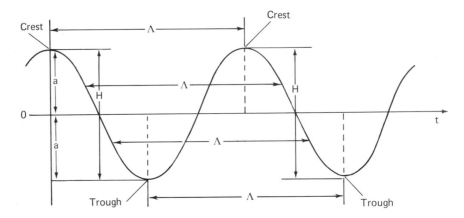

Figure 10.2. Parts of a simple sinusoidal wave. Note that the wave height is defined as twice the amplitude.

$$\kappa = \frac{2\pi}{\Lambda}, \qquad \omega = \frac{2\pi}{T}, \qquad C = \frac{\omega}{\kappa} \tag{10.2}$$

Wave motion is a bit subtle. You can take a length of rope and with a flip of the wrist start a wave moving down its length (Figure 10.3). The wave moves the entire length of the rope, but the rope has not moved any distance. Similarly with ocean waves, which travel along the surface while the water stays put. A cork floating on the surface bobs up and down as the waves pass, but it does not travel along the surface with the speed of the wave. It is necessary to distinguish between the movement of the wave (wave motion) and the movement of the water itself (particle motion).

A large portion of the observed characteristics of ocean waves can be accounted for by assuming that ocean waves are a combination of simple sinusoidal waves, such as shown in Figure 10.2. If one assumes that the wavelength is much longer than the wave height (a quite reasonable assumption for the ocean), it is possible to derive a relationship among wave speed, wave number, and depth of water (h) (see Appendix I):

$$C^2 = \frac{g}{\kappa} \tanh \kappa h \tag{10.3}$$

The hyperbolic tangent can be further simplified. When κh is small (less than 0.33), tanh κh is approximately κh. When κh is "large"

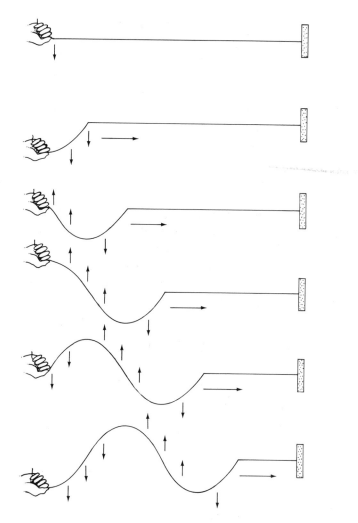

Figure 10.3. When you send a wave down a rope, the wave moves forward, but not the rope.

(greater than 1.5), κh is approximately unity. Substituting these approximations in Eq. (10.3) gives

$$C_s^2 = gh \tag{10.4}$$

$$C_d^2 = \frac{g}{\kappa} \tag{10.5}$$

where the subscripts s and d refer to shallow water waves and deep

water waves, respectively. The significance of the two approximations can be seen in Figure 10.4. For Eq. (10.4) to hold (to within
about 5%), the wavelength must be at least 20 times the depth of
water. For Eq. (10.5) to hold (again to within 5%), the wavelength
must be less than four times the depth of water. For wavelengths of
intermediate size, Eq. (10.3) must be used. Fortunately, most of the
waves of interest in the ocean can be characterized as either shallow
or deep water waves.

The waves described by Eqs. (10.4) and (10.5) are quite different
from one another. The speed of shallow water waves is controlled by
the depth of water; the deeper the water, the faster the wave. The
speed of deep water waves is independent of the depth. As long as the
water is sufficiently deep that the wavelength is no more than four
times the depth of water, the wave speed and other characteristics of
the wave are independent of the water depth. The speed of deep
water waves is, however, dependent upon the wavelength and period.
Combining Eqs. (10.5) and (10.1) gives

$$\left. \begin{aligned} C_d &= \frac{g}{2\pi} \ T \cong 1.5 \text{ m/s} \\[2em] \Lambda &= \frac{g}{2\pi} \ T^2 \cong 1.5 \ T^2 \text{ m} \end{aligned} \right\} \quad (10.6)$$

One way to visualize the relation of deep water waves to shallow

Function	Deep Water	Intermediate	Shallow Water
Phase Velocity, C	$\frac{g}{2\pi}T$	$\left[\frac{g}{k}\tanh kh\right]^{1/2}$	$[gh]^{1/2}$
Limits of Application	$\left[\frac{h}{\Lambda} > \frac{1}{4}\right]$		$\left[\frac{1}{20} > \frac{h}{\Lambda}\right]$

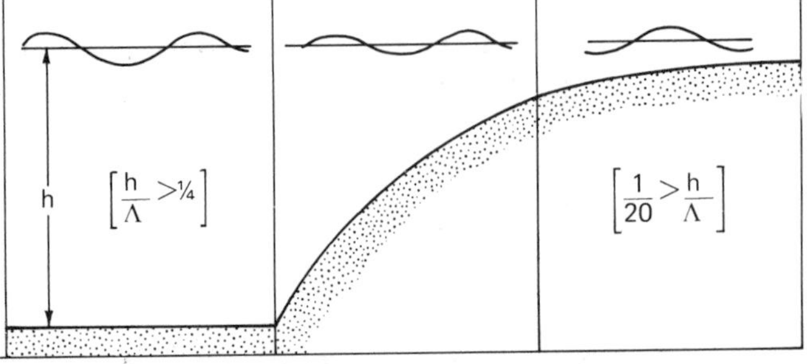

Figure 10.4. Limits of applicability of deep and shallow water waves.

water waves is to imagine what would happen to a deep water wave if its length and period were increased. Let the ocean be 4000 m deep and start with a 10-s wave. According to Eq. (10.6), it would have a wavelength of 150 m and a speed of 15 m/s, clearly a deep water wave. If the period were 20 s, the wavelength would be 600 m and the speed 30 m/s, still a deep water wave. It would continue to be a deep water wave until its period was about 100 s, at which time the wave speed would be 150 m/s, and its length about 15 km. By the time the period was 4 min, the wave would be a shallow water wave. It would be traveling at about 200 m/s and have a wavelength of nearly 50 km. If the period were 10 min, the wavelength would be about 120 km, but the speed would remain the same 200 m/s. The shallow water wave speed is the *maximum* speed of a surface wave. The relationship between speed, period, and water depth is shown in Figure 10.5 for the predominant surface gravity waves.

Particle Motion

Deep and shallow water waves differ in characteristics other than their phase velocity. Figure 10.6 shows the characteristic particle

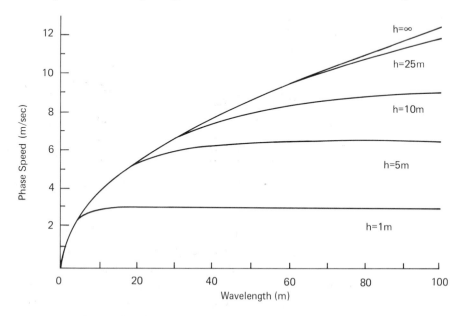

Figure 10.5. Wave speed as a function of water depth h and wavelength Λ. The water depth determines the limiting wave velocity, that of a shallow water wave.

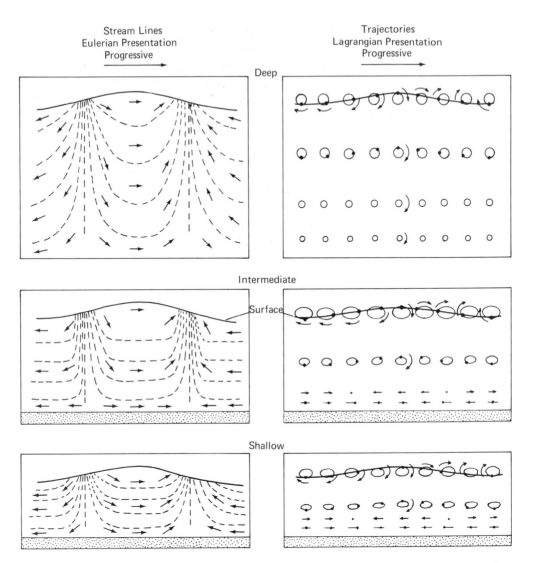

Figure 10.6. Streamlines and trajectories for deep, shallow, and intermediate waves.

motion for deep and shallow water waves. Consider the deep water case first. The individual water particles describe circles whose radius decreases with depth. At the surface the radius (r) is the same as the wave amplitude (a), and the particle velocity (V) is simply the circumference of the circle divided by the wave period. A pressure recorder just below the wave would record a pressure change (Δp) equal to the change in hydrostatic pressure as the wave passed by.

$$r = ae^{-\kappa Z}$$

$$V = \frac{2\pi}{T}ae^{-\kappa Z} \left.\begin{array}{l}\\ \\ \\ \\ \\ \end{array}\right\} \quad (10.7)$$

$$\Delta p = a\rho ge^{-\kappa Z}$$

The exponential relationship means these parameters decrease rapidly with depth. Where the depth Z is equal to half the wavelength, the radius, particle velocity, and recorded pressure differences are reduced to 4% of their surface values.

The shallow water case is more complex. The water particles describe ellipses. The radius along the minor axis is equal to the wave amplitude at the surface and decreases linearly with depth, until at the bottom the minor axis is zero and the motion is horizontal. The radius of the major axis is a function of water depth, wavelength, and amplitude and does not vary with depth. The pressure differential is simply the difference in hydrostatic pressure:

$$\Delta p = a\rho g \qquad\qquad (10.8)$$

It does not change with depth. Table 10.1 gives the exact formulation of the relevant parameters for deep and shallow water waves.

Table 10.1 Comparison of Relevant Parameters for Deep and Shallow Water Waves

	Deep	Shallow
Surface displacement, η	$a\cos(\kappa x - \omega t)$	$a\cos(\kappa x - \omega t)$
Phase speed, C	$\dfrac{gT}{2\pi}$	$\sqrt{gh}$
Particle velocity components, u, w	$u = \omega ae^{-\kappa z}\cos(\kappa x - \omega t)$	$u = \dfrac{\omega}{\kappa}\dfrac{a}{h}\cos(\kappa x - \omega t)$
	$w = \omega ae^{-\kappa z}\sin(\kappa x - \omega t)$	$w = \omega a\left(\dfrac{z}{h} + 1\right)\sin(\kappa x - \omega t)$
Pressure differential, Δp	$\rho gae^{-\kappa z}\cos(\kappa x - \omega t)$	$\rho ga\cos(\kappa x - \omega t)$
Semimajor axes A and semiminor axes B of particle path ellipse	$A = B = ae^{-\kappa z}$	$A = \dfrac{a}{\kappa h}$
		$B = a\dfrac{h + z}{h}$

A pressure recorder placed on the bottom is a common way to measure wave height in shallow water. As can be seen from Eq. (10.7) and (10.8), the interpretation of pressure records can be complex. If the wave period (and consequently the wavelength) is long enough to ensure that the passing wave is a shallow wave, then the pressure difference is simply related to wave height by Eq. (10.8). If, however, the period (and thus the wavelength) is sufficiently short that the wave is either intermediate or deep water, the pressure differential cannot be interpreted in terms of wave height by the hydrostatic relation of Eq. (10.7). It is, however, a fairly straightforward calculation to relate the observed pressure to the wave height, if the depth of water and the wave period are known.

Energy and Wave Dispersion

The energy in a wave (E) is equally divided between potential energy (the displacement of water from its equilibrium position) and kinetic energy (the energy associated with particle movement). For both deep and shallow water waves,

$$E = \tfrac{1}{8}\rho g H^2 \quad \text{or} \quad \tfrac{1}{2}\rho g a^2 \tag{10.9}$$

The units are in terms of energy per unit surface area. Thus the wave energy in a field 1-km square with the waves 1 m high is approximately 1.2×10^9 joules (J).

Waves whose speeds are frequency dependent are called *dispersive* waves. Deep water waves are dispersive waves; shallow water waves are not. Consider two wave trains of slightly different periods superimposed upon one another, as in Figure 10.7. The resulting envelope shows regions where the waves are in phase (where the wave energy is largest) separated by regions where the waves are out of phase (where the wave height and wave energy are minimum). If both wave trains traveled at the same speed, so would the resulting envelope. The maxima and minima, corresponding to the regions of in phase and out of phase, would also move at the same speed. However, if the wave trains move at different speeds, the regions where the two waves are in phase and out of phase will change (Figure 10.8). The regions of maximum wave height in the envelope travel at a slower speed than either of the individual wave trains.

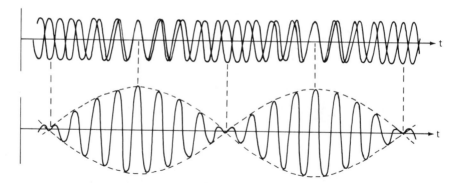

Figure 10.7. Two wave trains of similar amplitude but different periods cause beats. If the two wave trains travel at identical phase speeds, the "beats" will progress at the same speed.

Thus the wave energy is propagated more slowly than the phase velocity. The speed of the envelope is called the *group velocity* $\mathcal{V}$ and for dispersive waves with the characteristics of deep water waves it can be shown that

$$\mathcal{V} = \tfrac{1}{2} C_d \tag{10.10}$$

For shallow waves the group velocity is the same as the phase velocity:

$$\mathcal{V} = C_s \tag{10.11}$$

The derivation is straightforward once you conclude that the group speed of the envelope in Figure 10.8 is

$$\mathcal{V} = \frac{\Delta\omega}{\Delta\kappa} = \frac{\partial\omega}{\partial\kappa} \tag{10.12}$$

just as the phase speed of the individual wave is $C = \omega/\kappa$ (see Appendix I).

Those who have been to sea and have attempted to identify individual waves and watch their progress are continuously frustrated. The waves seem to disappear. The reason, of course, is that one is observing not individual wave trains of identical periods, but the envelope made up by adding together a series of wave trains of slightly different periods. Since the energy is propagated more slowly

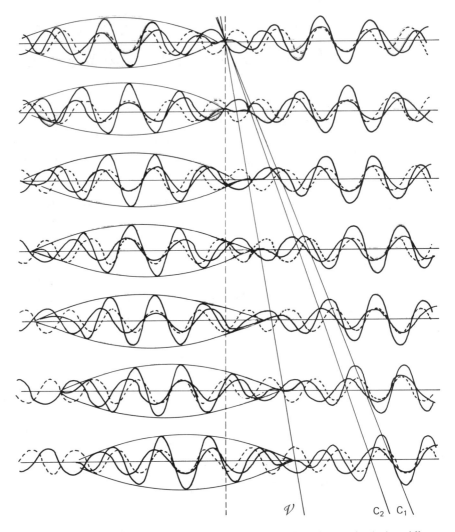

Figure 10.8. Two progressive wave trains of similar amplitude but different wavelength and wave speeds move in and out of phase. Individual wave crests can be tracked at wave speeds C_1 and C_2. Nodel point progresses at half the speed $\mathcal{V}$.

than the individual waves, what one sees is one wave appearing to slowly disappear, while another behind it seems to build. On the other hand, by the time a wave gets in shallow water, the group velocity and the phase velocity are identical. A surfer who picks a wave to ride need not worry that his wave will disappear, while the one behind builds.

The concept of wave energy moving at a speed different than the

Table 10.2 Wave Energy as a Function of Time and Distance from Wave Machine*

| Travel time in periods | Distance from wave generator in wave length† | | | | | | | | Total energy in the train ($\frac{1}{2}E$) |
	1	2	3	4	5	6	7	8	
1	$1/2$								1
2	$3/4$	$1/4$							2
3	$7/8$	$4/8$	$1/8$						3
4	$15/16$	$11/16$	$5/16$	$1/16$					4
5	$15/32$	$26/32$	$16/32$	$6/32$	$1/32$				4
6	$15/64$	$41/64$	$42/64$	$22/64$	$7/64$	$1/64$			4
7	$15/128$	$56/128$	$83/128$	$64/128$	$29/128$	$8/64$	$1/128$		4
8	$15/256$	$71/256$	$139/256$	$147/256$	$93/256$	$37/256$	$9/256$	$1/256$	4

*In part after Kinsman, B., 1965: *Wind Waves*, Prentice-Hall, Inc., Englewood Cliffs, N.J.
†Wave maker turned on to make four deep water waves and then turned off. The energy per wave is shown.

phase velocity is not an easy one. The following example has often been used to assist in the visualization of what occurs. It can be argued that the partition of energy in a deep water wave is equally divided between that which moves with the phase speed and that which does not move at all. The average therefore is an energy translation at half the phase speed.

Imagine a very long wave tank with a plunger at one end that produces simple harmonic waves. For convenience of notation we shall assume that each wave has an amount of energy E/2. As the waves move down the tank, each wave will translate half its energy forward at the phase speed and leave half behind. The resulting energy distribution after each of the first four waves is shown in Table 10.2. There is a small amount of energy that can be detected moving at the phase speed, but in compensation there is an increasing amount of energy building up near the wave generator.

Now turn the wave machine off after the generation of the fourth wave. The energy envelope spreads in both directions, but the total energy moves forward at a speed equal to half the phase velocity (see table 10-2).

Wave Formation and Capillary Waves

The most casual observer recognizes that wind causes surface waves on water. A flat calm sea or lake can become choppy within minutes after a strong wind rises. The longer and harder the wind

blows, the higher the sea. When the wind stops, the sea begins slowly
to subside. For years scientists have attempted to explain the mechanism by which the wind creates waves. However, after reviewing 50
years of effort in this field, Ursell could write in 1956, "Wind blowing
over a water surface generates waves in the water by physical processes which cannot be regarded as known." Considerable effort has
been expended on this problem since then, but Ursell's statement
continues to be true today.

It is not that plausible mechanisms have not been suggested, but
rather than none of them allows for the rapid buildup of surface waves
that has been observed. A 20-knot wind can develop waves a meter
high in a few hours. None of the mechanisms yet suggested can
account for such a rapid buildup. All recent attempts are based on
turbulent processes, transferring energy from the wind to the sea by a
coupling of the fluctuations in wind speed and local air pressure with
developing "roughness" profiles on the ocean.

The first effect of the wind blowing on a flat sea is to form small
surface tension or capillary waves. These waves are present,
superimposed on the larger waves and swells, whenever the wind
blows. They die out almost immediately when the wind stops. They
receive their name from the fact that, if the radius of curvature of the
air–water interface is only a few centimeters, then the force tending to
minimize the surface area (i.e., make the surface flat) is an appreciable fraction of the gravitational force. For a wave whose length is
less than 5 cm, Eq. (10.5) must be modified to include this additional
restoring force:

$$C^2 = \frac{g}{\kappa} + \frac{\kappa}{\rho}Y \qquad\qquad\qquad (10.13)$$

where Y is the surface tension. Surface tension waves, or capillary
waves, have rather interesting features. For one thing, the shorter
they are, the faster they go (Figure 10.9). For another, the group
velocity is greater than the phase velocity. Thus it appears that new
waves are continually building in front of the wave train while those
in the rear die out. The minimum speed for a wave from Eq. (10.13) is
22 cm/s. Its wavelength is 1.73 cm.

There are two reasons for at least briefly noting capillary waves.
The first is that, until we have a better idea that we have now about

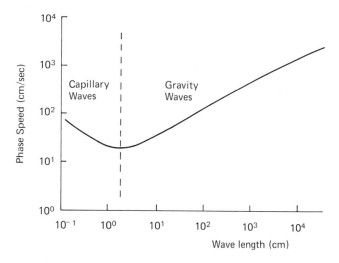

Figure 10.9. For waves less than 1 cm, speed is determined by surface tension. Waves longer than 10 cm are essentially gravity waves.

how the wind builds surface waves, the possible role of the capillary wave cannot be ignored. Capillary waves increase the microroughness of the sea surface and for this reason may play a significant role in wave formation.

The second reason is that capillary waves are responsible for the characteristic texture of the sea surface. Those occasional areas where the sea appears very smooth, the *surface slicks*, are areas without capillary waves. Slicks may be the result of the absence of sufficient wind to cause capillary waves to form, or they may be the result of oil or other material on the surface, which reduces the surface tension. Examples of the latter can occasionally be seen in rivers or estuaries, and even in the open ocean, whenever there is a convergence of surface currents. As the water comes together it must sink, but material floating on the surface remains along the convergence zone. Thus an area of convergence is also a concentrator of surface-floating material, such as oil and other organic matter. The surface tension is less, and the characteristics of capillary waves, if they form at all, are quite different. The passing swell has the same height and period within the slick area as outside it, but the optical effect resulting from the absence of capillary waves often makes the slick area appear smoother than the surrounding water.

Wave Spectrum and the Fully Developed Sea

The waves generated by the wind bear little resemblance to the idealized forms of Figure 10.7. The surface of the ocean appears as an almost random pattern of hills and valleys (Figure 10.10). Even when one examines the sea surface in one dimension, a simple wave pattern is not often evident (Figure 10.11). These complex surfaces, however, can be constructed by a process called spectral analysis by assuming that they are composed of a series of simple sinusoidal waves of varying amplitude (Figure 10.12). The square of the wave amplitude per frequency band is plotted along the vertical, the frequency (and period) along the horizontal. What Figure 10.12 shows is that a series of sinusoidal waves of the periods and amplitudes indicated, if added together in the manner of Figure 10.7, would give a wave profile with a similar statistical form to Figure 10.11. An analogous profile could

Figure 10.10. The surface of the sea does not resemble a simple sinusoidal wave pattern. (U.S. Navy photograph.)

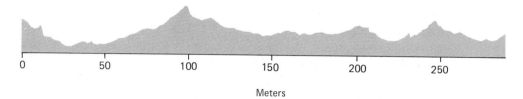

Meters

Figure 10.11. Wave profile can be obtained by drawing a line through Figure 10b and plotting wave height as a function of distance along the line.

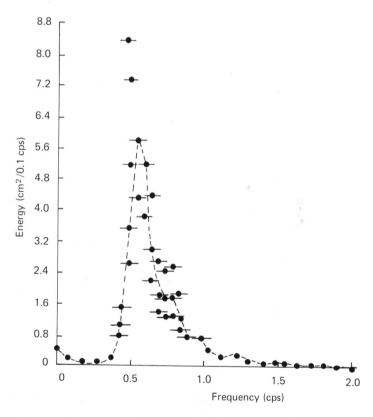

Figure 10.12. By averaging the square of the wave height over each frequency band from records such as Figure 10.11, one generates wave spectra plots such as this. (After Kinsman, B., 1965: *Wind Waves,* Prentice-Hall, Inc., Englewood Cliffs, N.J.)

be drawn from a profile perpendicular to the first, and from it a spectral analysis could be made. If these "two-dimensional" spectra are now added together, the resulting form is statistically similar to the two-dimensional picture of Figure 10.10.

The reason why the square of the amplitude and frequency are used in Figure 10.12 rather than amplitude and period is that amplitude squared is proportional to wave energy, and when plotted against frequency, the area under any portion of the curve is proportional to the amount of wave energy in that frequency (or period) interval.

As might be expected, wave spectra vary depending upon the strength of the wind and how long and over what distance it has been blowing. As the wind blows, waves are formed. The longer the wind blows, the higher the waves become. A wind of 12 knots blowing for a couple of hours will build waves that begin to break. At 50 knots the wind will blow the tops off the waves, and at 100 knots it is difficult to distinguish the "air–sea interface." For any given wind speed there is an equilibrium point at which the energy imparted to the waves by the wind equals the energy lost by the waves through breaking. This is called the *fully developed sea* and can be described by wave spectra (Figure 10.13). Most of the energy is found over a comparatively narrow frequency range. There is a shift in the characteristic period as the wind increases, from about 4 s for a 10-knot wind to about 16 s for a 40-knot wind. The average wave height, of course, increases with wind speed also.

Fully developed seas do not occur every time the wind blows. The wind must blow for a considerable time and over a considerable area. It takes a longer time to build a fully developed sea for a strong wind than for a light wind. A strong wind also requires a longer *fetch* over which to blow than does a light wind. If the fetch area is not long enough, the waves may propagate out of the generating area before they have had an opportunity to become fully developed. Figure 10.14 is a nomogram that relates minimum time and fetch necessary to build a fully developed sea.

Oceanographers often refer to average wave height, significant wave height, and the average of the highest 10% of the waves. For a given wave record, such as Figure 10.11, these statistics are generated by simply measuring the distance from trough to crest of each

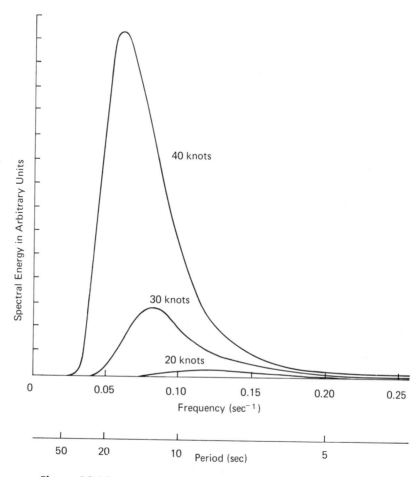

Figure 10.13. Idealized wave spectra for a fully developed sea for winds of 20, 30, and 40 knots.

wave in the record, and doing the necessary averaging. Significant wave height is defined as the average of the highest one third of the waves in the record. There is good evidence to suggest that when a careful observer at sea visually estimates the average sea height he is approximating the significant wave height, because he commonly eliminates the smaller perturbations of the sea surface from his visual average. The combination of theory and empirical evidence that generates spectra curves such as Figure 10.13 can be used to determine

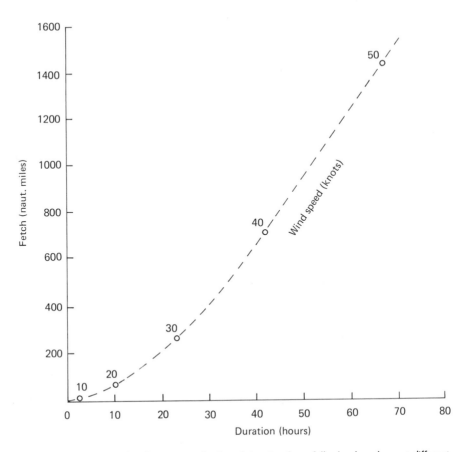

Figure 10.14. Minimum fetch and duration for a fully developed sea at different wind speeds. As shown in the example, a 30-knot wind must blow for nearly 24 h over a fetch of at least 290 miles to build a fully developed sea. If either the fetch or the duration is less than the minimum, the sea does not reach its steady-state conditions.

the other statistics of a fully developed sea (Figure 10.15). Fortunately, for those who go to sea the combined requirements of fetch and duration rarely occur for winds in excess of 40 knots. The conditions are found most frequently in the Antarctic Ocean.

Wave Propagation

Spectral analysis is more than a convenient shorthand for statistically summarizing complex wave patterns. It provides a means of relating what is observed in the ocean to the idealized waves. That

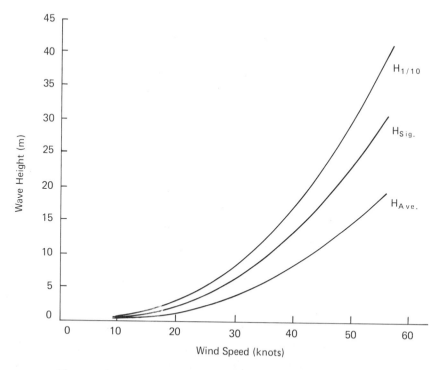

Figure 10.15. Predicted average, significant, and highest 10% wave heights for a fully developed sea at different wind speeds. (After Pierson, W. J., Jr., G. Neumann and R. W. James, 1955: "Practical Methods for Observing and Forecasting Ocean Waves," *U.S. Navy Hydrographic Office*, Pub., No. 603.)

this should be the case is not at all obvious, but because it is true we can make certain predictions about what happens to waves as they move out of the generating area. To a fairly good approximation, we may assume that the complex sea of Figure 10.10 does indeed consist of a combination of simple periodic waves, and that these waves propagate with the phase speeds and group velocities indicated for deep water waves. Making due allowance for a limited amount of frictional loss along the way, we can at any time add the various spectral components together and have a spectral analysis of the ocean surface that closely approximates the statistical properties of the observed sea surface. This mathematical trick works rather well, because there is apparently very little transfer of energy from one part of the spectrum to another; there is little *wave–wave* interaction.

To do such an analysis correctly requires more information than is usually available. As a result, a series of empirical relations has been derived. Assume that a storm 2000 km from the coast generates a fully developed sea comparable to Figure 10.13 for a 30-knot wind. As the storm leaves the generating area, a certain amount of diffraction occurs. As a rule of thumb, the ratio of the *significant wave height* decreases as

$$H = H \cos \theta \tag{10.14}$$

Thus at an angle of 30° from the direct downrange position of the storm, the significant wave height would be 86% of that observed in the direction of propagation (Figure 10.16). In the absence of any appreciable wind, all that one would observe beyond the fetch region is the slow undulation of the sea surface (swell) resulting from the storm.

Frictional losses do occur, and, as in the case of all waves, the shorter periods are attenuated more rapidly than the long. Thus there is a slight shift of the energy spectra toward longer periods.

All the significant wave components in Figure 10.14 have wavelengths that are short compared to the depth of the open ocean, which means that the energy in the different components of the spec-

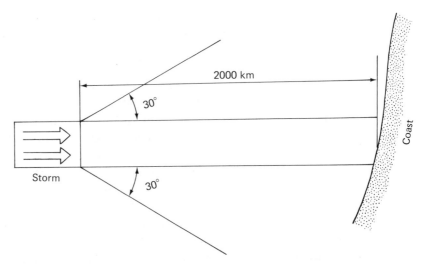

Figure 10.16. As waves move out of the generating area of the storm, they spread by diffraction. The swell height at an angle of 30° from a line downwind of the generating area would be about 85% of the directly downwind swell height.

trum is propagated at different speeds. The longer periods arrive first. The energy travels with the group velocity, *not* the phase velocity. Thus, according to Eqs. (10.6) and (10.10) a wave gauge near the shore would begin recording 18-s waves about 40 h after they left the generating area. The wave height would be measured in centimeters, and it would require a very good filter to see such small amplitudes superimposed on the usual background level of wave activity, which might be of the order of 1 m. However, 32 h after these early "forerunners," major swell from the storm would begin breaking on the beach.

An extreme example is the tracking of storms originating in Antarctica from Southern California. Munk and his colleagues have tracked storms originating more than 10,000 km away by observing the progressive arrival of dispersive waves, the first arrivals with periods of more than 20 s and heights measured in millimeters.

Refraction and Breakers

Except for the slight shift in the frequency spectra caused by selective attenuation of the higher frequencies, there is little significant change in the swell until it reaches the shore. As the water depth decreases, the waves cease to be deep water waves. Eventually, the water shoals sufficiently that they become shallow water waves and marked refraction can occur. Consider shallow water waves moving onto a beach with both an offshore spit and a canyon. Since the group velocity is $(gh)^{1/2}$, the waves adjacent to the spit are slowed relative to those in deeper water, while those in the canyon move faster. The wave crests are no longer parallel to the beach. Furthermore, as the waves slow up, the distance between crests decreases, because the period of the waves does not change; and if the speed decreases, the wavelength must decrease according to Eq. (10.1).

It is possible to draw orthogonals that are everywhere perpendicular to the wave crests. Assume for a moment that there are no frictional losses or diffraction during the brief interval of the waves' run toward shore. Under these circumstances the average amount of energy in the area between two crests and the two orthogonals remains constant. However, as the waves slow up, the distance between crests decreases and, therefore, the area decreases. Thus the wave height increases, since the average energy per unit area must increase. Superimposed on this foreshortening of the distance between crests is a further decrease of the distance between orthogonals over the spit and an increase of the distance between orthogonals over the canyon (Figure 10.17).

Another way to visualize what happens is in terms of energy flux. The flux of energy across a line between two orthogonals must be constant and is simply

$$\mathscr{V}E = \text{constant} \tag{10.15}$$

If the distance between orthogonals remains constant, E increases as the group velocity decreases. If the distance widens, the same flux of energy now moves across a wider front, which causes a decrease in the wave height unless the group velocity increases. Conversely, as the orthogonals converge, the wave height increases unless there is a decrease in the group velocity.

As the waves move closer and closer to shore, the wavelength decreases and the wave height increases until, when the ratio of wave height (twice the amplitude) to water depth becomes about 0.7, the

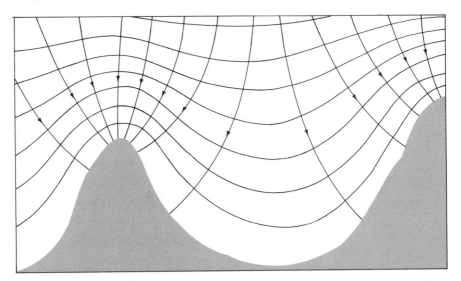

Figure 10.17. Waves slow down as they move into shallow water and as a consequence they are refracted toward shallow water. Since to a first approximation the flux of energy between two adjacent rays (the lines with the arrows) is constant, wave heights decrease over canyons and peak over ridges.

wave becomes unstable and breaks. The height of the breaking waves is related to the strength of the storm and its distance from shore. However, it should also be noted that local topography causes perturbations in the average breaker height. Waves breaking over the spit are higher than those breaking over the canyon.

Knowing the position and strength of the storm as well as the details of the local topography allows one to predict the time of arrival and approximate strength of the storm waves. Local topography plays an important role in determining the effect of a storm. Note, for example, that a storm whose predominate energy was in the 6-s period range would miss the lee side of the point of land in Figure 10.18. Waves with a 10-s period, however, would feel bottom sufficiently to be refracted around the point and on to the lee shore.

Finally, it should be remembered that storm waves travel great distances. The slow, heavy summer surf in Southern California can be traced to storms originating 10,000 km away in the Antarctic Ocean.

Longshore Currents and Rip Currents

Consider a long, straight shoreline with a gently sloping beach and a series of waves breaking offshore. Once the waves break there is an appreciable amount of water transported shoreward within the breaker zone. If the water level on shore is not to rise, this water must eventually find its way back through the surf zone. If the waves approach the beach at an angle, longshore currents are generated that run in the direction of the waves; if the waves break parallel to the beach, the longshore currents are generally symmetrical (Figure 10.19). In either case, small points of land, offshore bars, and the like, can complicate the simple picture of Figure 10.19.

The combination of breaking waves, wave-transmitted momentum and longshore currents results in variations in the mean water level and the resultant pressure field inside the surf zone. Generally, low-pressure areas are found where the breaking waves are lower than average; for example, over a canyon as in Figure 10.17. In these areas, rip currents form. Thus bottom topography and shoreline configuration influence the location of rip currents. In the absence of such unconformities, the location and spacing of rip currents are presumably related in some way to traveling "edge" waves along the

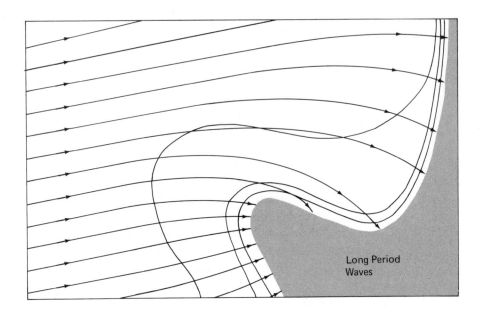

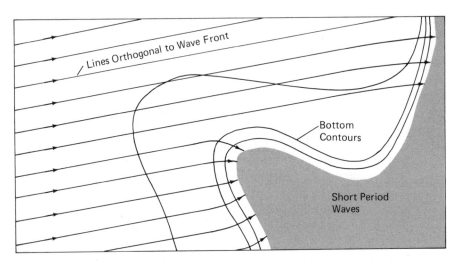

Figure 10.18. To forecast breakers on a beach, it is necessary to know the offshore bottom topography as well as the characteristics of the offshore waves. In the above example the lee of the point is sheltered from short-period waves, but longer-period waves will feel bottom sooner and be refracted around the headland.

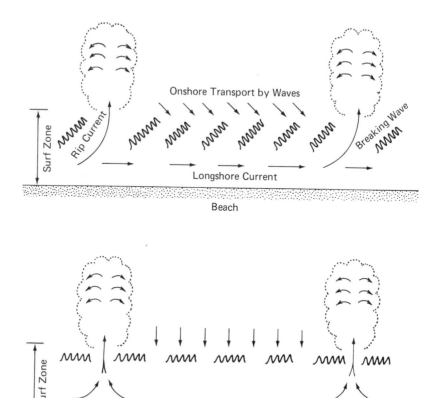

Figure 10.19. Breaking waves transport water to the beach. It must return through the surf zone. Usually, this surf-transported water runs along the beach, building to form discrete rip currents. Along a straight beach these rip currents are often equally spaced. (After Inman, D. L., 1971: "Nearshore Processes," in *Encyclopedia of Science & Technology*, McGraw Hill, Inc., New York, 9.)

beach (see Chapter 11). The dimensions of the cells between rip currents are usually two to eight times the width of the surf zone.

Rip currents are commonly referred to as *undertow* by swimmers. There is no evidence of strong subsurface currents in the surf zone other than the instantaneous backwash under a breaking wave on a steep beach. However, rip tides can be dangerous to the unwary swimmer. Rip tides of 3 knots have been measured carrying water seaward through the surf zone.

chapter 11

TIDES AND OTHER LONG-PERIOD WAVES

Most of the previous chapter was confined to a discussion of deep water waves (i.e., dispersive waves whose speed is a function of their period). This chapter will focus primarily on long-period, shallow water waves, nondispersive waves whose speed is a function of water depth and whose group velocity is the same as their phase velocity

$$C = \mathcal{V} = \sqrt{gh} \tag{11.1}$$

Tsunami

One of the most spectacular and devastating of the long-period waves is the tsunami. The fact that tsunamis are caused by earthquakes has been recognized for several hundred years, as has the additional puzzling, but happy, fact that not all underwater earthquakes cause tsunami. A tsunami wave is often referred to as a tidal wave in popular literature and in the scientific literature before

1950. Tsunami was suggested as an alternative to avoid confusion with tides, since tides and tsunami have little in common except that they are both shallow water waves. The effect of tsunami waves can be felt at great distances, and the damage and loss of life caused by waves more than 20 ft high has been considerable. A tsunami in 1896 killed an estimated 27,000 people on the coast of Japan. Tsunami warning systems exist, but perhaps only one in ten large underwater earthquakes causes noticeable damage, and the chances of destructive damage occurring in a given location are considerably less than that. Even the most timid hate to leave their homes when the false alarm rate is that high. Thus present systems usually require more evidence than notice of an earthquake before sounding the alarm.

Present evidence suggests that tsunami are directly coupled to crustal movement. A shallow-focus earthquake that causes major crustal movement of the ocean bottom will cause a tsunami. An equally strong earthquake that does not result in large crustal movement will not. A somewhat more remote possibility is that tsunami are triggered by major turbidity flows resulting from earthquakes.

At its origin the tsunami may appear as a single pulse, the leading edge of which travels with the speed of a shallow water wave. No one has successfully measured the wave shape in deep water remote from shore, but it is inferred that the initial pulse can be shaped such that energy need not be propagated equally in all directions. Thus even knowing that there was probable crustal movement is not sufficient to predict the region of major damage.

What can be predicted is the time of arrival. By knowing the location of the earthquake, which is routinely provided within a few minutes through a worldwide seismic network, and the depth of the ocean between the source and any possible target, the arrival time is easy to predict. In fact, the average depth of the Pacific was inferred in 1855 by A. D. Bache by noting the time of arrival of tsunami on various tide gauges.

The average depth of the oceans is about 4000 m, which means that a tsunami travels about 200 m/s (450 mph or 400 knots). It takes less than 5 h for a tsunami generated in the Gulf of Alaska to reach Hawaii. As it moves into shallow water, the tsunami slows down. At a depth of 100 m, its speed is about 30 m/s; at 50 m it is reduced to 22 m/s.

In the absence of friction, energy flux is conserved; that is, $\mathscr{V}\mathrm{E}$ is constant [Eq. (10.15)]. Remembering that wave energy per unit surface area is proportional to the square of wave height, and that group velocity and wavelength are proportional to the square root of the depth, it is easy to show that the depth is inversely proportional to the fourth power of the wave height. Thus a wave that is 15 m high when it breaks in 20 m of water has a wave height slightly more than 5 m when the water depth is 4000 m. If one assumes a characteristic period of 10 min, the wavelength in deep water is about 120 km, resulting in a slope of the sea surface of 1 in 30,000. It is little wonder that tsunami are not seen by ships at sea

It is less clear why tsunami cause more damage in some localities than in others. Distance from the source, local refraction effects, and focusing of the source pulse are all important; but it is also thought that a wide continental shelf serves both as a wave reflector (sending much of the energy back across the ocean) and as an absorber of tsunami energy through friction along the bottom. Shallow shelves can also trap wave energy. Whatever the reasons, incidents of significant tsunami damage are rare in regions with wide continental shelves.

Seiches and Other Trapped Waves

The slow oscillation of the water level of a lake or harbor is called a *seiche*. A seiche is a shallow water standing wave, and perhaps the best known example of a seiche is the sloshing wave one can generate in a bathtub. There are a variety of ways in which a seiche may be excited in a natural body of water. One of the most common in a lake is by the passage of a storm in which the wind raises the water level at one end of the lake. When the wind dies and the stress of the wind is removed, the lake surface begins to oscillate.

To understand the physics of a seiche, first consider the motion of water particles in a standing wave. At the nodes the water motion is entirely horizontal. At the antinodes it is vertical. Elsewhere, it has both horizontal and vertical components (Figure 11.1). Consider next a simple channel closed at both ends in which the water surface can be made to oscillate. The simplest mode is that shown in Figure 11.2. Note that the oscillation has the characteristic of a wave, whose wavelength is twice the length of the channel. The boundary conditions require a wave antinode, vertical movement only. The wave is

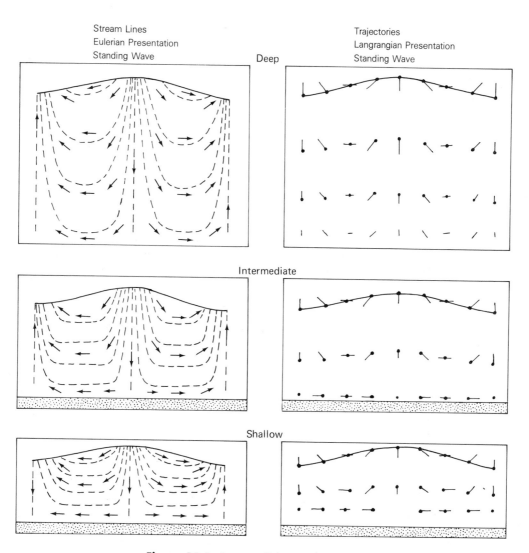

Stream Lines
Eulerian Presentation
Standing Wave

Deep

Trajectories
Langrangian Presentation
Standing Wave

Intermediate

Shallow

Figure 11.1. In a standing wave the particle motion is horizontal at the nodal points and vertical at the antinodes. The figure shows a crest in the middle and troughs at the ends. A half period later, the water will crest at both ends and there will be a trough in the middle. At the nodes (which are halfway between crest and trough) there is no vertical movement.

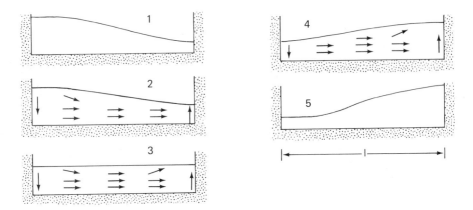

Figure 11.2. Since there can be no flow through the boundary, a seiche in a closed basin requires an antinode at each end of the basin. For the fundamental mode this means the basin is a half-wavelength.

clearly a shallow water wave. A rigorous derivation is beyond the scope of this book, but perhaps the following will suffice:

$$C = \frac{\Lambda}{T} = \frac{2l}{T} = \sqrt{gh}$$

$$T = \frac{2l}{\sqrt{gh}}$$

(11.2)

Thus, if one knows the depth and length of the channel, (ℓ), one can determine the period of oscillation. Equation (11.2) and Figure 11.2 are based on a single node midway in the channel. It is possible to have additional nodes, and it can be easily shown following the logic of Eq. (11-2) that in general

$$T = \frac{2l}{n\sqrt{gh}}$$

(11.3)

where n is the number of nodes.

Next consider a channel open at one end to a level ocean (Figure 11.3). Here the boundaries require a node at the channel opening and an antinode at the end. Thus the channel length is a quarter of a wavelength:

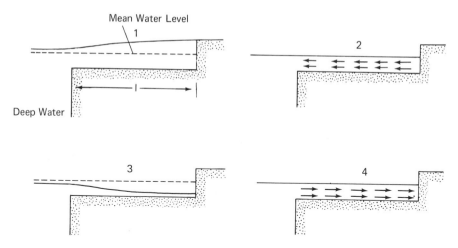

Figure 11.3. Boundary conditions for a seiche in a narrow basin open at one end (for example a bay) require a nodal point where the bay connects with the ocean and an antinode at the closed end. For the fundamental mode this means that the basin is a quarter-wavelength long.

$$C = \frac{\Lambda}{T} = \frac{4l}{T} = \sqrt{gh}$$

$$T = \frac{4l}{\sqrt{gh}}$$

$$(11.4)$$

For an open channel with more than one node

$$T = \frac{4l}{n\sqrt{gh}} \qquad (11.5)$$

where $n = 1, 3, 5, \ldots$

Knowing the natural period of a channel, lake, or harbor is of some importance, and there is a considerable body of literature on the subject of seiches. The simple formulation of Eqs. (11.3) and (11.5) requires modification when considering problems with variable bottom topography, harbors where the width is an appreciable fraction of the length, or where the expected seiche may have a wave height such that the small-amplitude assumption may be invalid. Occasionally, small boat harbors have been built that have a natural period close to that of the prevailing swell or other natural forcing functions, with the result that the harbor is in a continual state of oscillation and not of much use for anchoring boats.

Seiche-like oscillations have been observed between islands and in other semienclosed areas, but not all such changes in sea level can be explained by seiches. Waves can be trapped on the continental shelf and run along the coastline. These are called *edge waves*, and their characteristics are determined by the slope and width of the shelf, as well as the wave number.

Storm Surges

Occasionally, hurricanes or storms cause currents and sea-level changes much greater than one would predict by simple considerations of momentum transfer by wind stress. Such phenomena are generally characterized under the classification of storm surges. The passages of storm surges have been well recorded. Much of the damage caused by hurricanes is a result of high water level associated with these waves. Although the primary storm surge itself is associated with the passage of the storm, this can be followed by edge waves, seiches, or other trapped waves. If one of these after waves should occur at high tide, the damage can be greater than that caused by the original storm surge, even though the storm center has passed. The art of predicting these occurrences is not far advanced. A common characteristic of storm surges is that they depend in some way upon resonant phenomena. Two such mechanisms have been suggested.

Consider a hurricane moving over the ocean. The pressure in the center of the hurricane could be of the order of 100 mbars less than in the surrounding areas. By the inverse barometer effect (see last section in this chapter), this would result in sea level being about 1 m higher in the center of the hurricane. If the hurricane travels at a speed of $(gh)^{1/2}$, then coupling resonance occurs; that is, the hurricane is moving at the same speed as a shallow water wave generated by the atmospheric pressure differential. Shallow water waves move in the deep ocean at about 200 m/s (almost 400 knots), much swifter than the fastest-moving storm. On the continental shelf the shallow water wave velocity is of the order of 40–60 knots. Only rarely do storms move that fast, and luckily for those concerned about the resulting shore damage, such storms are generally rapidly "filling"; that is, pressure in the low-pressure center is increased and the storm is weakened.

A second form of resonant coupling by a storm can occur when the wind shifts in concert with the natural seiche period of the underlying water. As a cyclonic storm, such as a hurricane, moves over a point, there can be a 180° shift in the wind. For example, as a hurricane moves northeastward over a semienclosed bay, the wind can build up from the southwest for a 12-h period and then shift to northeast as the eye passes; then the wind blows for 12 h from the northeast. If the natural period of the bay is 24 h, the seiche generated from the passing hurricane would be enhanced.

Passing storms can generate storm surges even when the resonance is not exact. Storm surges have been recorded along the east coast that have been tied to hurricanes moving up the Atlantic coast. The arrival of these storm surges need not coincide with the period of maximum wind. One of the most devastating and best documented storm surges in modern times occurred in the North Sea in the winter of 1953. Sea level rose 3 m above normal. It has been shown that there was good resonant coupling both with respect to the speed of the storm and the period of the wind shift.

Internal Waves

Ocean waves are not limited to surface waves. Waves may be found throughout a stratified fluid. As a simple, and idealized, example, consider two fluids of density ρ' and ρ'', as in Figure 11.4. Waves may form along the interface. The general solution is

$$C^2 = \frac{(\rho'' - \rho')\,(g/\kappa)}{\rho''\,\coth\,\kappa h'' + \rho'\,\coth\,\kappa h'} \tag{11.6}$$

Although this appears to be rather complicated, note that if $\rho'' \gg \rho'$, as is the case of the density of water compared to the density of air, Eq. (11.6) reduces to

$$C^2 = \frac{g}{\kappa}\,\tanh\,\kappa h'' \tag{11.7}$$

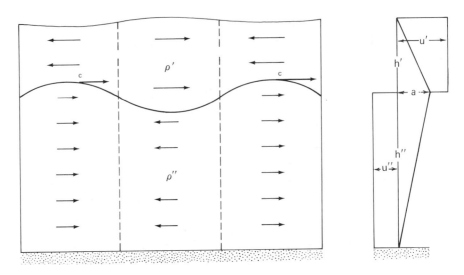

Figure 11.4. Shallow water internal wave in a simple two-layer ocean.

which is identical with Eq. (10.3). We might consider surface waves as a special class of two-layer internal waves in which the density of the surface layer is so much less than that of the lower layer that it may be ignored.

Solutions of Eq. (11.6) can be found corresponding to the deep and shallow water solutions of Eqs. (10.4) and (10.5). If the wavelength were sufficiently long that both top and bottom layers were "shallow," then

$$C^2 = \frac{gh''h'}{h'' + h'}\left(1 - \frac{\rho'}{\rho''}\right) \tag{11.8}$$

which is similar to the shallow water wave solution. In the ocean a typical value of $1 - (\rho'/\rho'')$ would be 0.002 or less. The speed of the internal wave would be only a few percent of the surface wave. Small wave tanks have been made consisting of two layers of fluid of slightly different densities. The resulting waves along the interface appear to move in slow motion.

Although consideration of a two layer ocean may be of some help in conceptualizing internal waves, the real ocean is not a two-layer ocean. It is continuously stratified, and internal waves occur any-where within the water column. The frequency range of internal waves is limited by two parameters. They cannot have a higher fre-quency than the Brunt–Väisälä frequency (see Chapter 1) or a lower frequency than the inertial frequency (Chapter 7). Except in the re-gion of a sharp thermocline, internal wave periods are measured in hours rather than minutes, and nowhere in the ocean are there inter-nal waves with periods measured in seconds.

Our knowledge of internal waves is fragmentary at best. A number of mechanisms have been suggested for generating internal waves, including ocean currents flowing over complex bottom topog-raphy and the change in atmospheric pressure of a moving storm system; but all such suggestions must be classified as "plausible hypotheses" for the present, pending better observational evidence. There is good evidence to suggest that much of the internal wave energy is centered around inertial and tidal frequency. Horizontally propagating internal waves can have a number of nodes and antinodes between the surface and the bottom, and the waves can propagate vertically or at an angle, as well as horizontally. There is also some evidence to suggest that internal wave amplitudes are higher near the continental shelf than in the open ocean (Figure 11.5). Breaking inter-nal waves have been observed, and it has been suggested that this mechanism plays an important role in ocean mixing and in turbulent decay (Figure 9.14).

Tidal Forces

Tides are caused by the difference in gravitational forces result-ing from the change of position of the sun and the moon relative to points on the earth's surface. The force of gravity is proportional to the product of the masses of the two objects and inversely propor-tional to the square of the distance between them.

For simplicity, consider the effect of the moon on the two points *a* and *b* on the earth's surface along the earth–moon axis (Figure 11.6). There is a balance between the earth–moon gravitational attrac-tion and the centrifugal acceleration of both with respect to a common axis of rotation. However, although the centrifugal force is the same

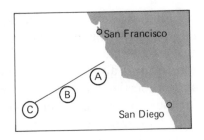

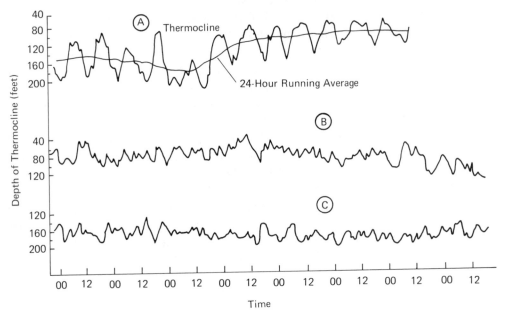

Figure 11.5. At station A, close to the coast, the thermocline rose and fell nearly synchronously with the tide. At stations farther from shore, the strong internal wave tidal component was less evident. (After Reid, J. L., Jr., 1956: "Observations of Internal Tides in October 1950," *Trans. Amer. Geophy. Un.*, 37.)

for all points on the earth, there are slight differences in the gravitational attraction. For example, the gravitational attraction of the moon M on a particle m at point a is

$$F_a = G\frac{m\mathrm{M}}{(P - R)^2} \tag{11.9}$$

The centrifugal force at a is

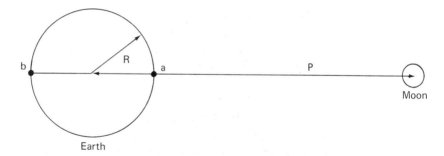

Figure 11.6. Relationship between the moon and points *a* and *b* on the earth used in Eqs. (11.9) to (11.14).

$$F_c = G \frac{mM}{P^2} \tag{11.10}$$

The imbalance at point *a* is

$$F_a - F_c = G \frac{mM}{(P - R)^2} - G \frac{mM}{P^2} \tag{11.11}$$

which reduces to

$$F_a - F_c = \frac{2PR - R^2}{P^4 - 2P^3R + P^2R^2} \, GmM \tag{11.12}$$

Since $P \cong 60R$,

$$F_a - F_c \cong + \left(\frac{2R}{P^3} \right) GmM \tag{11.13}$$

Going through the same argument for point *b* gives

$$F_b - F_c = - \left(\frac{2R}{P^3} \right) GmM \tag{11.14}$$

Thus we have two equal and opposite forces at points *a* and *b* whose force is proportional to the mass of the moon and inversely proportional to the third power of the distance.

Although the trigonometry is a bit more complicated for points other than *a* and *b*, there is nothing more complicated in principle in the calculation of the tidal force for any point on the surface of the earth. Figure 11.7 indicates schematically the nature of the vector forces.

The mass of the sun is about 2.5×10^7 times that of the moon, but it is also 400 times farther away. Substituting exact numbers for mass and distance into Eqs. (11.13) and (11.14) shows that tidal forces produced by the moon are slightly more than twice those of the sun.

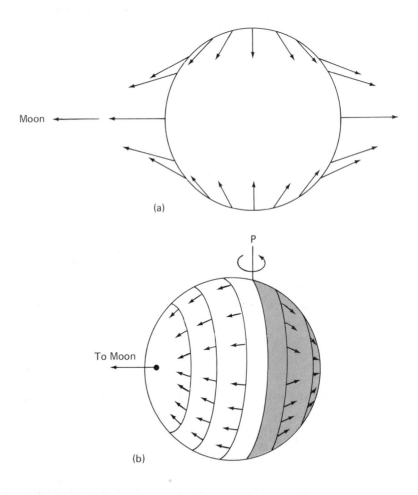

Figure 11.7. (a) Distribution of the total tide-generating forces for a cross section of the earth; (b) the horizontal component of the tide-generating force at the surface of the earth.

The tidal forces are very weak compared to the earth's gravitational force. They are only about 1 part in 9 million. It is the horizontal component of these forces (those acting along the surface of the earth) that causes the water to move (Figure 11.7b).

Equilibrium and Dynamic Theory of Tides

The simplest concept of tides is the equilibrium tide. With an earth covered by water, there would be a rise and fall of sea level corresponding to the tidal forces that would exactly balance the pressure-gradient forces resulting from the sloping sea surface (Figure 11.8). The heights of this equilibrium tide can be calculated. The maximum lunar tide would be 55 cm, the solar tide 24 cm, combining to produce a maximum of 79 cm at full moon and new moon when the sun and moon are nearly in line with the earth. The concept of an equilibrium tide, which was first proposed by Newton, does account for the fact that the maximum tides (spring tides) occur at full and new moon and that neap tides occur at the quadratures. Although tidal heights are generally larger than those predicted by equilibrium theory, the ratio of height of spring tide to those of neap tide is about as predicted. However, the observed tides are not in proper phase. For example, spring tide should occur everywhere on earth the time either the new or full moon is directly overhead, and it does not.

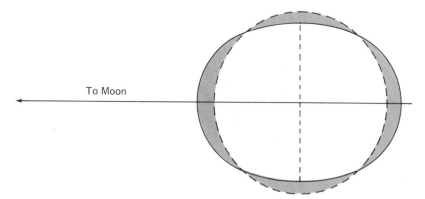

Figure 11.8. Distribution of tides under equilibrium theory assuming on earth completely covered with water. The dotted line represents the surface of the earth without tidal forces; the solid line represents the shape of the surface with tidal forces applied. The distortion is much exaggerated. For the moon above it would be about 55 cm.

Laplace was the first to suggest a dynamic theory of tides in which the tides are considered as waves driven by the periodic fluctuation of the tidal forces. The problem would be difficult enough if the earth were everywhere covered with water to a uniform depth. No one has yet succeeded in solving the problem for the real oceans. The essence of the problem is relating a known periodic driving force to a series of interconnecting basins, each with its set of natural frequencies to which it can respond, and each with its own frictional characteristics.

Consider the following example. Assume that there is a narrow channel girdling the earth at the equator. We further simplify the problem by ignoring the sun. The tidal forces of the moon produce a wave that travels around the earth with two crests and two troughs, similar to that of the equilibrium tide in Figure 11.8. Each crest will stay under the earth–moon axis, and thus will move halfway round the earth in about 12 h (more precisely, 12.4 h allowing for the orbit of the moon about the earth). These are shallow water waves and the wave velocity is 450 m/s. Since the speed of a shallow water wave is $\sqrt{gh}$, we can calculate that the depth of the channel should be about 21 km.

If the channel were indeed 21 km, the free period of the wave would be equal to the period of the driving force, and we would have what is called resonance. Under the influence of the periodic driving force, the tidal waves in the channel would grow larger and larger. Their size would be limited only by the internal friction in the water and along the sides and bottom of the channel. It is probably just as well that there is not such an earth-girdling channel in resonance with the periodic driving forces. One of the few places where there is a basin whose natural frequency does correspond to about 12 h is the Bay of Fundy, whose 15-m tides are the highest in the world.

The problem of relating the response of a machine with one natural period to that of a driving force of a different period is a classical one in mechanics. Consider what happens when a weighted spring with a natural period of 6 s is driven by a vibrator with a 4-s period. The spring will oscillate at 4 s, but the amplitude of oscillation will be low. If the vibrator period is 5 s, the spring will also oscillate at 5 s, and the amplitude will be somewhat larger. It will be even larger if the vibrator's period is 5.5 s, and, of course, it will be in resonance at a period of 6 s (Figure 11.9).

The average depth of the ocean is 4000 m. The natural period of a channel that deep corresponds to a free shallow water wave with a

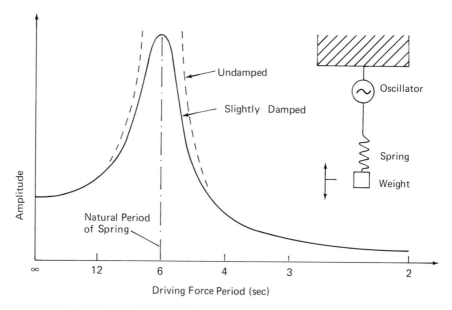

Figure 11.9. A spring will oscillate with the period of the periodic driving force. The amplitude of the oscillation increases toward resonance when the period of the oscillator is the same as the natural period of the spring.

speed of about 200 m/s. The 12.4-h periodic driving force of the tides will produce a wave with a 12.4-h period, traveling at about 450 m/s, but its amplitude will be only 6 cm. Figure 11.10 shows the observed heights of the tide around the world.

Ocean Tides

From a combination of partial solutions to the theoretical problem outlined in the last section, together with data from the thousands of tide gauges found on land, it has been possible to piece together a picture of the major features of the tidal waves in the ocean. Figure 11.11 shows the semidiurnal tide in the Atlantic and in the Black Sea. The solid lines represent times of high tide referred to the passage of the moon by the meridian of Greenwich. Note that the North Atlantic is dominated by a counterclockwise rotary tide, connected to a progressive tide that moves up from the South Atlantic. The pivot point in a rotary tide is called an *amphidromic point*. The Black Sea tide on the other hand is a simple clockwise rotary tide with the amphidromic point near the center.

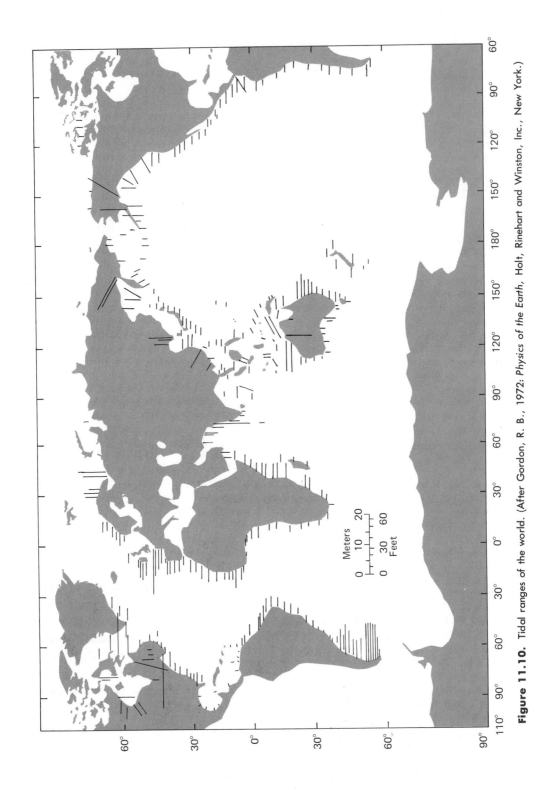

Figure 11.10. Tidal ranges of the world. (After Gordon, R. B., 1972: *Physics of the Earth*, Holt, Rinehart and Winston, Inc., New York.)

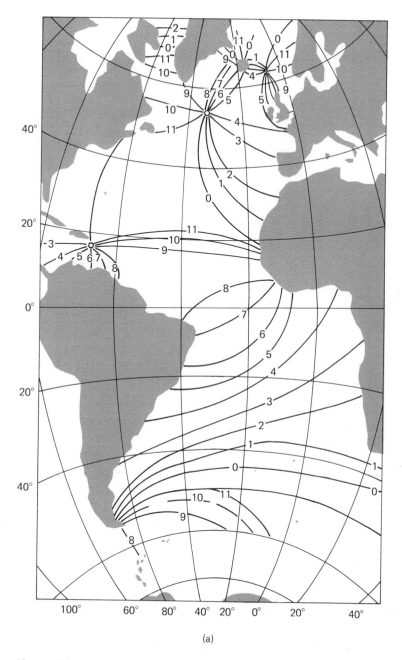

(a)

Figure 11.11. (a) Semidiurnal M_2 tide in the Atlantic is a combination of a counterclockwise rotating tide in the North Atlantic in combination with a progressive tide that runs up and down the South Atlantic; (b)—on the following page—in the Black Sea the rotary tide is clockwise. The pivot points in the Black Sea and the North Atlantic are amphidromic points with zero tidal amplitude. The numbers refer to the hour of high tide.

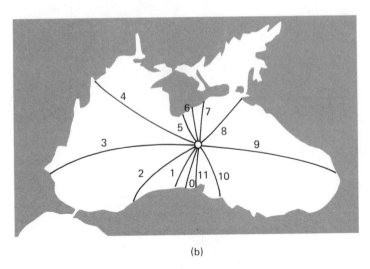

(b)

Figure 11.11b.

Although tidal charts such as Figure 11.11 have been drawn for many ocean basins, the information on which they are based is not as substantive as one might expect when one contemplates the data from 10,000 tide gauges. Nearly all tide gauges are from the edges of the ocean. There are insufficient islands in the ocean, and not all these have tide gauges or are properly located, to give the coverage necessary to test charts such as Figure 11.11. A few tide gauges are now being placed in the open ocean (very accurate pressure gauges sitting on the bottom). These, combined with better knowledge of the details of the bottom topography, coupled with large high-speed computers to do the very lengthy computations necessary for any theoretical solution applied to the real ocean, give hope of improving our knowledge of tides significantly.

Tidal Prediction and Other Changes In Sea Level

In spite of the apparently unsatisfactory state of our knowledge of tides, tidal prediction is a highly advanced art, and reliable tide tables can be prepared many years in advance. They can, that is, if tidal data have been collected in the past. The position and movement of the sun and moon are known in great detail. From this information the tidal forces can be determined. The simplest presentation of these forces is in a series of periodic functions. The principal harmonics are

Table 11.1 Principal Tidal Harmonic Components

Name of Partial Tides	Symbol	Speed (degrees per mean solar hour)	Period in solar hours	Coefficient ratio $M_2 = 100$
Semidiurnal components				
Principal lunar	M_2	28.98410	12.42	100.0
Principal solar	S_2	30.00000	12.00	46.6
Larger lunar elliptic	N_2	28.43973	12.66	19.2
Lunisolar semidiurnal	K_2	30.08214	11.97	12.7
Larger solar elliptic	T_2	29.95893	12.01	2.7
Smaller lunar elliptic	L_2	29.52848	12.19	2.8
Lunar elliptic second order	$2N_2$	27.89535	12.91	2.5
Larger lunar evectional	ν_2	28.51258	12.63	3.6
Smaller lunar evectional	λ_3	29.45563	12.22	0.7
Variational	μ_2	27.96821	12.87	3.1
Diurnal components				
Lunisolar diurnal	K_1	15.04107	23.93	58.4
Principal lunar diurnal	O_1	13.94304	25.82	41.5
Principal solar diurnal	P_1	14.95893	24.07	19.4
Larger lunar elliptic	Q_1	13.39866	26.87	7.9
Smaller lunar elliptic	M_1	14.49205	24.84	3.3
Small lunar elliptic	J_1	15.58544	23.10	3.3
Long-period components				
Lunar fortnightly	M_f	1.09803	327.86	17.2
Lunar monthly	M_m	0.54437	661.30	9.1
Solar semiannual	S_{sa}	0.08214	2191.43	8.0

shown in Table 11.1. The use of eight terms is sufficient to include all terms that contribute as much as 10%.

There are enough variations in tidal height due to local effects such as wind and atmospheric pressure to ignore all but the most important terms in developing tide tables. Those listed in Table 11.1 are sufficient for nearly all purposes.

What is done is to collect a tidal record for about a year at the location for which tidal prediction tables are desired. By harmonic analysis, the relative importance of the different periodic tidal terms is determined. An example for a single day is shown in Figure 11.12. Once the relative amplitudes and phase relationship of the different constituents have been determined, it is only a matter of making the calculations. Tide-prediction machines (essentially analog computers) were built as early as 1876. More recently, standard digital computers have taken over the task of tide predictions.

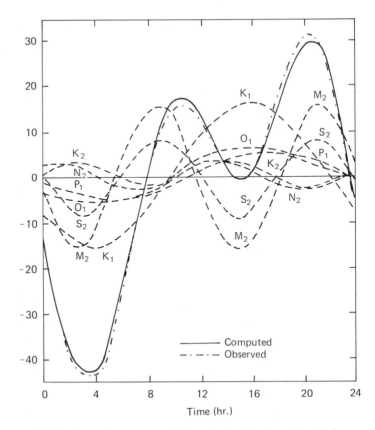

Figure 11.12. By applying proper weighting functions to the various tidal components (M_2, S_2 K_1, K_2, etc.) and adding in proper phase, one can compute the expected tidal curve, which in the example shown agrees well with the observed tide. (After Defant, A., 1961: *Physical Oceanography*, Pergamon Press, New York.)

A tide-prediction table is useful in predicting sea level, but on any given day there may be variations from the predicted values. Such variations are more marked in estuaries or other semienclosed bodies than on the open coast. Local wind effects can cause variations in predicted tidal heights. Variations of tens of centimeters are not uncommon in bays and estuaries. Other changes are a result of barometric pressure. A reduction of 1 mbar (about 0.03 in. of mercury) in atmospheric pressure results in a 1-cm rise in sea level, the *inverse barometer effect*. The heating of the water in summer causes about a 20-cm rise in sea level simply by thermal expansion, since warm water has a larger volume than cold water.

Finally, it should be noted that sea-level changes can result from shifts in major currents. Hamon has reported differences in sea level of a more than $\frac{1}{2}$ m over a several-month period, resulting from a shift in the direction and intensity of the currents flowing along the east coast of Australia.

Tidal Currents

Generally, tidal currents become stronger as one approaches the coast, and they play an increasingly important role in the local circulation. For example, tidal currents are at least 10 times stronger than nontidal currents in a partially mixed estuary. On the other hand, tidal currents in the open ocean are generally believed to be of the order of 2–5 cm/s, although there is considerable evidence for intermittent internal wave motion of tidal period generating periodic currents several times larger.

Surface tidal currents are rotary; that is, a water particle would follow the path of an ellipse during one complete tidal cycle (Figure 11.13). Predicting the amplitude, the ratio of major to minor axis, and direction of rotation requires detailed knowledge of the tidal wave itself. However, as one approaches the shoreline from the sea, one can safely assume that the major axis of the ellipse becomes more or less parallel to the shore; that is, the tidal currents are predominantly parallel to the shore.

It is also easy to show that tidal currents tend to increase as the water depth decreases. The maximum horizontal particle speed of a shallow water wave is (Table 10.1)

$$u = C \frac{a}{h} \tag{11.15}$$

and since $C = (gh)^{1/2}$,

$$u = a \left(\frac{g}{h} \right)^{1/2} \tag{11.16}$$

Thus, if the wave amplitude (a) is constant, the tidal current increases as the water depth decreases.

Finally, it should be noted that stronger than normal tidal currents can be expected in those areas of the world where the natural

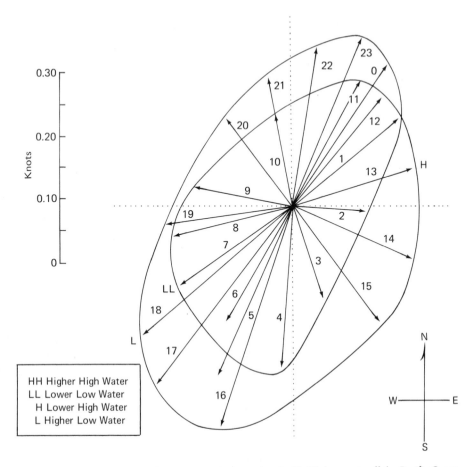

Figure 11.13. Rotary motion associated with tidal currents off the Pacific Coast. The strength of the tidal currents varies with the stage of the tide.

period of the basin is close to tidal frequency, so that the tidal forces generate a standing wave of tidal period. The Bay of Fundy is the best known example. For all these reasons (dynamics of estuarine circulation, particle velocity enhancement in shallow water, and generation of seiches of tidal period), the tidal currents are an increasingly large component of the total ocean current kinetic energy in nearshore regions. However, predicting the exact role of tidal currents in any given area is usually not possible without measurements. Once a series of tidal-current measurements has been made and related to tidal height, future tidal currents can be predicted in an analogous manner based on tide tables.

chapter 12

SOUND AND OPTICS

    The ocean is relatively opaque to all forms of electromagnetic radiation from very long radio waves to the short ultraviolet. What small transmission window exists is in the visible spectrum; but even there the transmission is poor compared to that of the atmosphere, since less than 1% of the energy reaches 100 m in the clearest ocean waters. Any "penetration" of the ocean by radio waves occurs only along the ocean surface; but for very low frequency radio transmission (where the wavelengths are of the order of kilometers), even a penetration of a fraction of a wavelength is of some importance to submerged submarines listening for radio messages from their home base. If one is to transmit information or "see" through the ocean space for any appreciable distance, he cannot depend upon electromagnetic radiation. The ocean, on the other hand, is much more transparent to sound transmission than is the atmosphere. A few pounds of TNT detonated off Hawaii can be heard by an underwater microphone (a hydrophone) off San Francisco. Even more dramatic, a depth charge exploded near Australia was monitored by hydrophones near its antipode off Bermuda.

Some Definitions Used in Underwater Sound

Because of its importance in a number of areas, not the least of which is tracking submerged submarines, it is probable that more scientists have spent more time with more elaborate equipment studying underwater sound than any other physical phenomenon in the ocean. As a result, our understanding of underwater sound is relatively good. The following discussion is phrased primarily in terms of the sonar equation, in which all quantities are measured in a nondimensional unit, the decibel. We shall start with a few definitions.

Intensity: If the oceans were truly incompressible (as we assume in most of our discussion of circulation, waves, and other hydrodynamic processes), there would be no sound. Sound is produced by the periodic condensation and rarefaction of the fluid. This results in a small pressure fluctuation Δp about the normal hydrostatic pressure. The intensity of the sound I is proportional to the square of this pressure fluctuation:

$$I = \frac{(\Delta p)^2}{\rho c}$$

where ρ is average density and c is the velocity of sound. Intensity has units of energy/area–time. It measures the flux of sound energy past a point, the sound energy traveling with the speed of sound (see Chapter 4 for a discussion of other types of flux).

In the discussion of sound, we shall adopt the convention that $\Delta p \equiv p$:

$$I = \frac{p^2}{\rho c} \tag{12.1}$$

Sound pressure varies over wide ranges, from less than 10^{-4} dynes/cm², lower than the ear can hear, to greater than 10^6 dynes/cm², which is equal to atmospheric pressure. Sound energy that intense at the sea surface would result in the formation of bubbles (cavitation).

Sound Level and Decibels: Sound intensity is usually discussed in terms of a nondimensional quantity called a decibel (dB), which requires defining a reference sound pressure. For architectural acoustics the reference is $p_{ref} = 2 \times 10^{-4}$ dynes/cm², which is the threshold of the human ear. In underwater sound the reference pressure is 1 dyne/cm².

Sound level in decibels is 10 times the log to the base 10 of the ratio of the measured sound intensity to the reference intensity. Consider the following example: a sound pressure of 1200 dynes/cm² is recorded:

$$\frac{I}{I_{ref}} = \frac{p^2/\rho c}{p^2_{ref}/\rho c} = \frac{p^2}{p^2_{ref}} = \frac{(1.2 \times 10^3)^2}{1^2} = 1.44 \times 10^6$$

Taking the log to the base 10,

$$\log \frac{I}{I_{ref}} = \log 1.44 \times 10^6 = 6.16 \text{ bels}$$

$$10 \log \frac{I}{I_{ref}} = 61.6 \text{ dB}$$

For a sound intensity with a sound pressure of 1200 dynes/cm², the *sound level* (**L**) is 62 dB.

Sound Velocity: The velocity of sound in the ocean varies with pressure, temperature, and salinity. A reasonable average value is 1500 m/s or 5000 ft/s, which is four to five times faster than the speed of sound in the atmosphere. Sound velocity increases with increasing temperature, salinity, and pressure. As can be seen in Figure 1.13, the relationship between sound velocity and temperature and salinity is nonlinear. Such approximations as 3 m/s/°C and 1.3 m/s/unit of salinity must be used with caution. Sound velocity is, however, a linear function of pressure. Since hydrostatic pressure is nearly linear with depth, one can write

$$\frac{\Delta c}{\Delta Z} \cong \frac{1.7 \text{ m/s}}{100 \text{ m}} \cong 0.017 \text{ s}^{-1}$$

The most commonly accepted relationship for sound velocity in seawater is by Wilson. It may be written

$$c = 1449 + 4.6T - 0.055T^2 + 0.0003T^3$$
$$+ (1.39 - 0.012T)(S - 35) + 0.017Z \qquad (12.2)$$

where T is temperature in centigrade, S is salinity, and Z is depth in decibars. Tables for calculating sound velocity are given in Appendix II.

Frequency: Sound pitch is usually measured in terms of frequency. The human ear can hear sounds as high as 15–18,000 hertz (Hz) and as low as 20–50 Hz, where 1 Hz is 1 cycle/s. Assuming a sound velocity of 1500 m/s, and remembering that wave velocity is simply wavelength divided by wave period, or wavelength times wave frequency, typical wavelengths for underwater sound are as follows:

Frequency	10^2	10^3	10^4	10^5
Wavelength	15 m	1.5 m	15 cm	1.5 cm

Sonar Equation

A number of useful concepts can be illustrated in terms of the sonar equation. Furthermore, the sonar equation itself is a useful tool for describing underwater sound processes. The usual units of the sonar equation are decibels, and the number of terms in the equation is determined in part by the degree of exactness desired. We shall discuss only the simplest terms.

Spherical Spreading: Consider the following problem. We have a sound source of known intensity (sound level), and we wish to echo range off a school of deep-swimming fish to determine their position and something about their composition. How sensitive a receiver will be required to receive the echo? The sound pressure in our source is 1000 dynes/cm², or a sound level (L_0) of 60 dB. We shall assume that it is a very simple sound source (e.g., a small charge of TNT), and that the sound radiates from this point source in all directions. We shall further assume that the sound velocity in this particular ocean is constant and, as a first approximation, that there is no frictional loss; that is, there is no *sound absorption*.

Remembering that intensity is energy flux and that this sound energy is being propagated at the same speed in all directions, it follows that, if there is no attenuation, the total flux is constant through a series of spherical shells subtended from a solid angle θ (Figure 12.1):

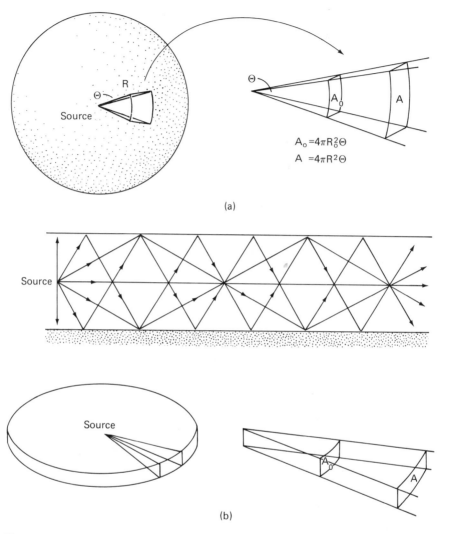

$A_0 = 4\pi R_0^2 \Theta$

$A = 4\pi R^2 \Theta$

(a)

(b)

Figure 12.1. (a) Ratio of the area A to A_0 increases as the square of the ratio of the distance R to R_0 for the case of spherical spreading, and (b) as the ratio of the first power of the distance R to R_0 for cylindrical spreading.

$$4\pi \boldsymbol{\theta} \mathbf{R}_0^2 I_0 = 4\pi \boldsymbol{\theta} \mathbf{R}^2 I \qquad (12.3)$$

$$\frac{I_0}{I} = \frac{\mathbf{R}^2}{\mathbf{R}_0^2}$$

In discussing the sound level of our source, it is necessary to be precise in indicating where the sound pressure was measured. Sound level of a source is always defined in terms of a unit distance from the source. In this case the unit distance (R_0) is 1 m. Taking 10 times the log to the base 10 of both sides,

$$10 \log I_0 = 20 \log R - 20 \log R_0 + 10 \log I \qquad (12.4)$$
$$L_0 = 20 \log R + L$$

since the log of 1 is zero. At a distance of 1000 m,

$$20 \log R = 60$$

Thus the sound level at 1000 m is 0 dB, which is the sound pressure equivalent of 1 dyne/cm².

Some of the sound that reaches the school of fish is reflected. The reflected sound also spreads spherically. If the fish were a perfect reflector and of "unit cross-sectional area," the relationship between the sound level generated at the source (L_0) and the sound level of the reflected sound L would be simply

$$L_0 = 20 \log R + 20 \log R + L \qquad (12.5)$$
$$L = L_0 - 40 \log R$$

For $L_0 = 60$ dB and the distance $R = 1000$ m, $L = -60$ dB, which is a sound pressure of 10^{-3} dynes/cm².

Target strength: Of course, some targets are better reflectors than others. Generally, a large target reflects more sound than a small target. Some objects are shaped to focus reflected sound; others disperse it. Equally important are the reflective characteristics of the target. To allow for these differences, a target strength term (in decibels) is added to the sonar equation

$$L = L_0 - 40 \log R + S \qquad (12.6)$$

where the larger the value of S, the better the target as a reflector of sound. Target strengths are generally determined empirically through measurement.

Cylindrical spreading: For sound transmission over long distances, the spreading cannot be spherical because the ocean is not infinitely deep. Consider Figure 12.1b. If the bottom and the surface are perfect reflectors, the sound spreads as a cylinder, not as a sphere. In the case of cylindrical spreading, it can be easily demonstrated that the intensity decreases as the first power of the distance, not the second power of the distance as in spherical spreading. In the case of the sonar equation, one- and two-way transmission losses are

one-way $\mathbf{L} = \mathbf{L_0} - 10 \log \mathbf{R}$

two-way $\mathbf{L} = \mathbf{L_0} - 20 \log \mathbf{R} + \mathbf{S}$

$$(12.7)$$

Cylindrical and spherical spreading are idealized examples. As might be expected, the situation in the ocean is more complex. However, as a general rule, if the transmission distances are short compared to the depth of the ocean, the transmission loss is approximately as the square of the distance (spherical spreading). If the range is long compared to the ocean depth, the sound is reflected from the surface and the ocean floor (Figure 12.1b). The transmission loss approaches the first power of the distance (cylindrical spreading).

Absorption and scattering: Although the ocean is reasonably "transparent" to sound waves, it is not completely so. Some absorption does take place because of viscosity and other properties of the medium. As in the case of light waves, the attenuation varies with frequency and follows Beers' law, where b is the attenuation coefficient:

$$\frac{dI}{d\mathbf{R}} = -bI$$

$$I = I_0 e^{-b\mathbf{R}}$$

$$(12.8)$$

In terms of decibels,

$$10 \log I = 10 \log I_0 - 10 \, b\mathbf{R} \log e$$

$$\mathbf{L} = \mathbf{L_0} - a\mathbf{R}$$

when $a = 10b \log e$ and is in units of decibels per unit length. Equations (12.6) and (12.7) for two-way transmission now become

$$L = L_0 - 40 \log R - 2 aR + S$$

$$L = L_0 - 20 \log R - 2 aR + S \qquad (12.9)$$

Over wide frequency ranges the absorption coefficient increases with the square of the frequency. For the range of 5–100 kilohertz (kHz), it is approximately

$$a = 1.5 \times 10^{-8} f^2 \text{ dB/km}$$

For f = 10,000 Hz,

$$a = 1.5 \text{ dB/km}$$

For freshwater, the absorption is even less than in seawater. It is

$$a = 2.08 \times 10^{-10} f^2 \text{ dB/km}$$

The additional loss in saltwater is related to an ionic relaxation phenomenon in the $MgSO_4$ salts, which absorbs sound energy.

By comparison, the absorption for 10,000-Hz sound in the atmosphere is about 160 dB/km for slightly moist air (relative humidity 40%). Absorption increases with increasing moisture content. In contrast to sound, the absorption of 30-kHz radio signals in seawater is about 1 dB/30 cm, and is higher at higher frequencies.

Sound energy is also "scattered." The small temperature and salinity inhomogeneities in the ocean (i.e., microstructure) cause reflection, diffraction, and refraction of the sound energy, which are generally lumped under the term *scattering*. Scattering is also caused by plankton and air bubbles. Because of scattering, the loss of sound energy between the source and the target is higher than one would calculate on the basis of absorption alone. The combination of absorption plus scattering is generally referred to as *sound attenuation*.

Ambient noise: The essence of any listening problem is that the received sound (**L**) must be loud enough to hear. The hearing can be by the human ear or by the most highly sensitive listening devices. Many times the nature of the hearing problem is not the level of the received sound but the background noise over which it must be heard. A conversation between two people can be carried on in whispers in an empty room; the same two may have to shout to be heard at a party. The necessary sound level must be sufficiently strong that it can be perceived against the background noise. In its simplest form, we can indicate this requirement as

$$L > N \tag{12.10}$$

where **N** is the noise level in decibels at the receiver.

In fact, Eq. (12.10) is much oversimplified. A careful listener can often understand the essentials of the conversation of his friend, even if the sound level of the conversation is less than the noise level of the party. Similarly, it is often possible to discern the line marking the ocean bottom on a recording echo sounder, even when the noise level on the chart would appear to be higher than the reflected signal from the bottom. In both cases, the information of the "ordered message" is sufficiently redundant that it is only necessary to occasionally have the required sound level higher than the noise level to piece together the message, even in the presence of an *average* noise level that is higher than the *average* sound level.

In the ocean the ambient noise level comes from several sources. The wind on the water causes sound, as do currents flowing over the bottom, breaking waves, and even rainfall. Additionally, there is biological noise. At certain times of the year the mating call of the toadfish drowns out most other sounds in certain nearshore regions of the U.S. east coast. Porpoises have their own echo-location devices (usually of high frequency and therefore rapidly attenuated). Certain whales on the other hand use low-frequency sounds, presumably for communication, and these sounds travel long distances.

Finally, of course, there are man-made sounds, such as the propellor noise of ships and the acoustic signaling devices of echo sounders and other instruments. Some years ago a study of the ambient noise level caused by winds and currents had to be abandoned in Narragansett Bay because of the large increase in motor boat traffic, which saturated the low-level recording devices. As might be expected, the ambient noise level varies with frequency; and since low-frequency sound is absorbed less, one might expect somewhat higher noise levels at lower frequencies, assuming, of course, that the sound source radiates sound energy over a broad range of frequencies. Figure 12.2 gives some indication of the sources and power of the ambient noise level in the ocean.

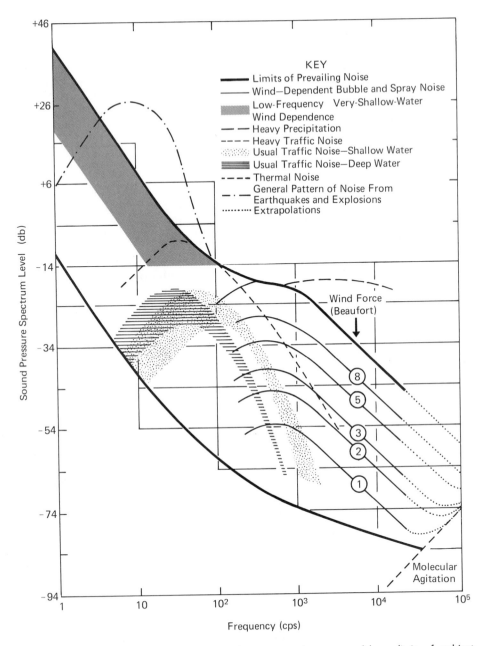

Figure 12.2. Heavy black lines suggest the upper and lower limits of ambient noise in the ocean (earthquakes, explosives, and heavy precipitations fall outside the envelope). The effect of wind-generated noise increases with wind speed (After Wenz, G. M., 1962: "Acoustic Ambient Noise in the Ocean: Spectra and Sources," *J. Acoust. Soc. Amer.*, 34.)

Refraction and Reflection

Since the velocity of sound is not constant in the ocean, sound rays do not travel in straight lines. The basic relationship is given by Snell's law, which relates the angle of incidence between layers of different sound velocity:

$$\frac{c_1}{c_2} = \frac{\cos \vartheta_1}{\cos \vartheta_2} \tag{12.11}$$

From the basic relationship one can derive a "critical angle":

$$\cos \vartheta_c = \frac{c_1}{c_2} \tag{12.12}$$

For angles less than ϑ_c, no refraction into the higher-velocity layer can occur (Figure 12.3).

It can also be shown that, in the case of a constant sound velocity gradient dc/dz, the sound ray describes an arc of a circle whose radius is

$$r_c = \frac{c_0}{(dc/dz) \cos \vartheta_0} \tag{12.13}$$

where ϑ_0 is the angle of the sound ray with the horizontal at velocity c_0 (Figure 12.3).

Sound rays are bent toward lower velocity. In the case of a typical thermocline, sound rays can be refracted such that there is a shadow zone in which no direct rays pass (Figure 12.4). Hovering under a shallow thermocline was a common way for submarines to attempt to avoid detection from surface ships in World War II; since the laws of physics have not changed in the intervening years, it is probably still a valid tactic. However, more powerful sonars that bounce sound off the bottom make this tactic less of a sure thing.

The effect of pressure tends to increase sound velocity with depth. Except for a few polar areas where the temperature is isothermal, the effect of the temperature structure of the ocean is to tend to

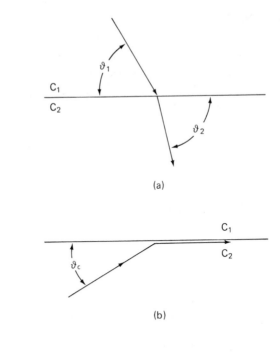

(a)

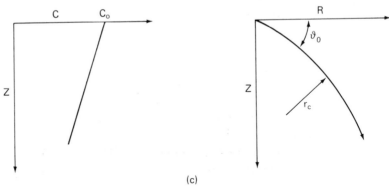

(b)

(c)

Figure 12.3. (a) Refraction as defined by Snell's law, Eq. (12.11); (b) the critical angle, less than which sound is trapped in the low velocity layer, Eq. (12.12); (c) the radius of curvature r_c is defined in terms of the initial angle ϑ_0 and the velocity gradient, Eq. (12.13).

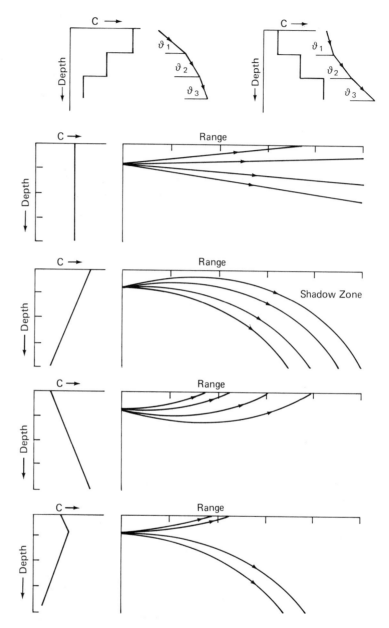

Figure 12.4. Under conditions of sound velocity decreasing with depth, sound rays are refracted downward; they are refracted upward with the gradient reversed. Under certain conditions, "shadow zones" develop.

decrease sound velocity with depth, with the effect much more pronounced near the surface than at depth. The combination of temperature and pressure effects typically results in a sound velocity curve
with a minimum (Figure 12.5). This minimum is called the *deep sound
channel*. Since sound rays are always refracted toward the lower
velocity, a sound velocity minimum serves to channel the sound (Figure 12.6). Sound channel transmission losses follow the general law of
cylindrical spreading.

The deep sound channel is a continuous feature in the ocean. It
extends from near the surface in polar latitudes to nearly 2000 m off
the coast of Portugal. A more typical depth is 700 m. In the case of the
very long transmissions referred to in the beginning of this chapter,

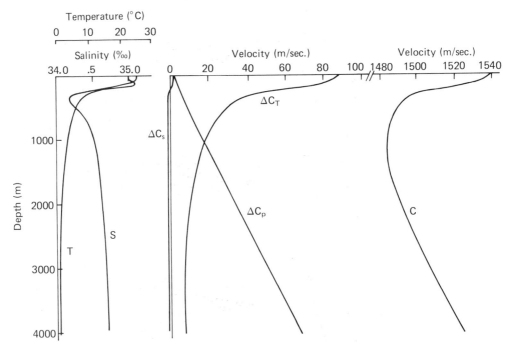

Figure 12.5. For the range of values one finds in the ocean, salinity has relatively
little influence on the sound velocity curve. The temperature effect is most pronounced in the thermocline. In the surface mixed layer and below the thermocline,
sound velocity increases with depth because of the pressure effect. The first panel
shows a typical temperature and salinity profile for the North Pacific. The second
panel indicates the effect of temperature, salinity, and pressure on the sound velocity
(information from sound velocity tables in Appendix II). The third panel shows the
actual sound velocity curves with a broad sound channel between 500 and 1200 m.

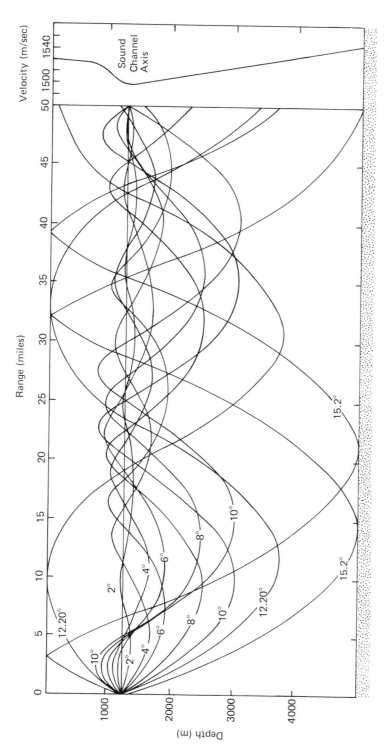

Figure 12.6. Typical sound ray patterns resulting from a deep sound channel. In the ideal case, as drawn, the sound source is at the axis of the sound channel, and all sound energy initially trapped in the sound channel stays trapped. In the real ocean, small perturbations in the sound velocity field resulting from temperature and salinity microstructure cause some of the energy to be scattered out of the sound channel. The angles refer to the direction of the initial ray with the sound channel axis as the ray leaves the source.

both the sound source and the hydrophones were near the deep sound channel axis. Since whale sounds have been recorded over long distances, it has been speculated that they, too, make use of the excellent transmission qualities of the sound channel.

The extent to which a sound ray is reflected from the surface or the ocean bottom depends upon the characteristics of the two media. The important parameter is the product of density and sound velocity. The greater the ρc difference between the two media, the more sound that is reflected and the less that is refracted. For the limiting case of ρc being the same in both layers, there is no reflection. In the ocean there is almost total reflection at the air–water interface, and much less reflection at the water–sediment interface. For given ρc characteristics, the amount reflected varies with the angle of incidence. The smaller the angle (the more tangential the ray to the interface), the more sound energy that is reflected. The relationship (see Figure 12.7 for an explanation of the notation) is

$$\frac{I_r}{I_i} = \frac{\rho_2 c_2 \sin \vartheta_i - \rho_1 c_1 \sin \vartheta_t}{\rho_2 c_2 \sin \vartheta_i + \rho_1 c_1 \sin \vartheta_t} \tag{12.14}$$

Uses of Underwater Sound

Underwater sound is used for communication, tracking submarines, measuring the depth of the ocean and the thickness of the sediments below, finding schools of fish, and in a variety of

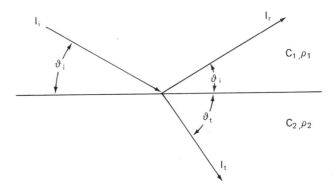

Figure 12.7. Ratio of energy refracted across a density interface to that reflected back is a function of the acoustical impedance (ρc) and the angle of incidence (ϑ_i), Eq. (12.14).

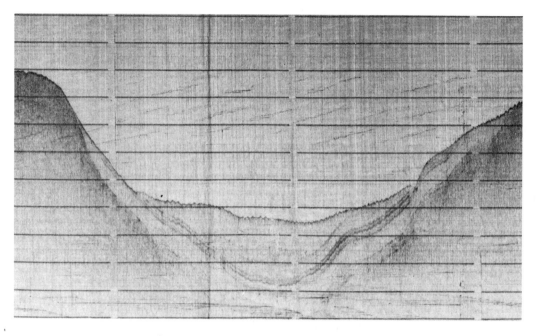

Figure 12.8. High-energy sound sources can show details of the subbottom continental shelf, as in this figure of ponded sediment in a valley. (From Emery, K. O.,: "The Continental Shelves," *Scientific American*, September, 1969. Record by U.S. Geological Survey, with permission.)

oceanographic instruments. The decision as to what frequency to use depends upon a tradeoff between low attenuation at low frequencies and better target definition (and often cheaper construction costs) at higher frequencies. A simple rule of thumb for echo ranging is that the target should be large compared to the wavelength. For this reason, ships looking for shallow, thin gravel deposits or similar layering prefer to accept the high attenuation loss of relatively high frequency echo sounders in order to get the detailed structuring of layers only a few meters thick (Figure 12.8). On the other hand, those prospecting for oil need deep penetration, which in turn requires low-frequency sound. Loss of detail is compensated by the ability to discern structures several hundred meters below the bottom (Figure 12.9). Echo ranging (sonar) is used in ways comparable to radar, and again the choice of frequency is governed by the size of the target and its distance. For echo location of nearby fish schools, one can accept high-frequency sonars, which give sufficient detail to tell the experienced observer the type of fish (Figure 12.10). The goal of military sonars is to find very

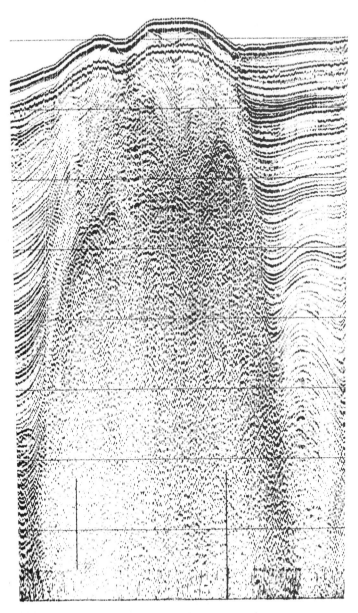

Figure 12.9. The "salt dome" thrusting its way upward through the Continental Shelf can be easily seen. This particular structure is in the Gulf of Mexico. Subsequent drilling produced oil. (From Emery, K. O.: "The Continental Shelves," *Scientific American*, September, 1969. Record by Teledyne Exploration Company, with permission.)

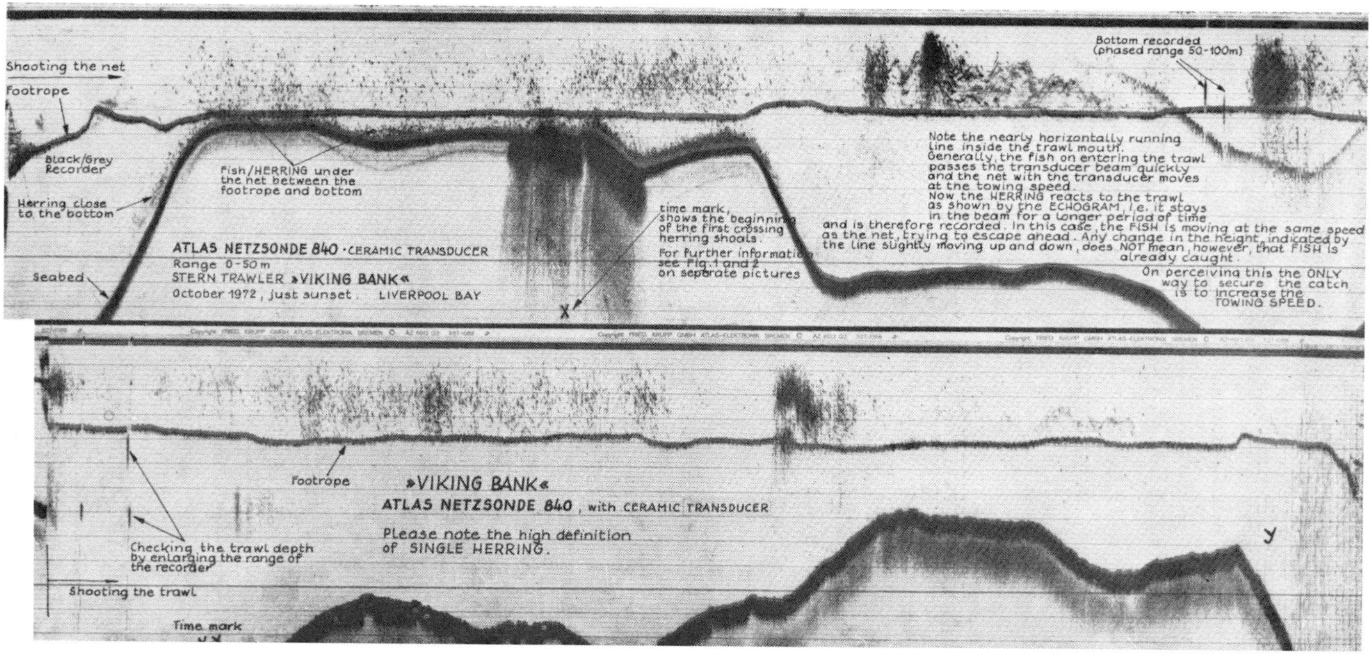
Shooting the net
Footrope
Black/Grey Recorder
Herring close to the bottom
Seabed
Fish/HERRING under the net between the footrope and bottom
ATLAS NETZSONDE 840 · CERAMIC TRANSDUCER
Range 0-50 m
STERN TRAWLER »VIKING BANK«
October 1972, just sunset. LIVERPOOL BAY
time mark, shows the beginning of the first crossing herring shoals.
For further information see Fig. 1 and 2 on separate pictures
Bottom recorded (phased range 50-100m)
Note the nearly horizontally running line inside the trawl mouth. Generally, the fish on entering the trawl passes the transducer beam quickly and the net with the transducer moves at the towing speed.
Now the HERRING reacts to the trawl as shown by the ECHOGRAM, i.e. it stays in the beam for a longer period of time and is therefore recorded. In this case, the FISH is moving at the same speed as the net, trying to escape ahead. Any change in the height, indicated by the line slightly moving up and down, does NOT mean, however, that FISH is already caught.
On perceiving this the ONLY way to secure the catch is to increase the TOWING SPEED.
Footrope
»VIKING BANK«
ATLAS NETZSONDE 840, with CERAMIC TRANSDUCER
Please note the high definition of SINGLE HERRING.
Checking the trawl depth by enlarging the range of the recorder
Shooting the trawl
Time mark
X
Y

Figure 12.10. The experienced fisherman can use an echo sounder to tell how far his nets are fishing off the bottom and whether he is catching fish. The wavelike pattern in the echogram is produced by the moving up and down of the boat (and echo sounder) with the surface waves (With permission of Krupp Atlas Elektronik.)

large objects such as submarines at long distances; thus relatively large low frequency sources are used. Sound is used for underwater "beacons," for tracking floating instruments, as well as for relaying data from such instruments. The difficulty of using sound energy rather than electromagnetic energy should not be underestimated. The equipment is usually bulkier, the power requirements greater, and the options and flexibility considerably less with sound than with radio or radar; but in principle there is an acoustic analog for every technological use of the electromagnetic spectrum in the atmosphere.

Underwater Optics

Because the ocean is so relatively opaque to electromagnetic radiation, the field of underwater optics has been studied relatively little. Much of what is known is in relation to the penetration of sunlight, a subject of some importance to those interested in oceanic photosynthesis, and much of the fundamental observational data (including the first demonstration that polarized light can be observed in the ocean) has been contributed by biologists as a consequence of their investigation of various biological problems.

As the light enters the sea, it is refracted. The law of refraction, Snell's law, relates the angle of the incoming beam to that of the transmitted beam by relative index of refraction. The index of refraction between air and water is about 1.33, and increases slightly with temperature and salinity (Figure 12.11). One consequence of refraction is that, to a swimmer underwater, the sun always appears higher than it is. Refraction can also result in distortion of underwater photography, since there is water on one side of the camera lens and air on the other. It should be noted that Snell's law applies to a flat surface. Wave action can cause deviations of as much as 15% in the expected direction of maximum radiation. It can also cause focusing effects; flashes of light can occur that may be dangerously high for an underwater swimmer looking upward.

Sunlight that penetrates the surface is both absorbed and scattered. The former is simply the transfer of radiant energy into heat; the scattering can be molecular or by particle matter. The amount of absorption and scattering is wavelength dependent. As in the atmosphere, the molecular scattering increases with the inverse fourth

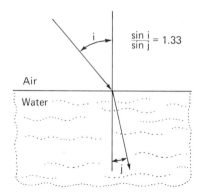

$$\frac{\sin i}{\sin j} = 1.33$$

Air

Water

Figure 12.11. Change in the angle of transmission as the sun's rays enter the water is a function of the air–water index of refraction.

power of the wavelength; that is, the blue end of the spectrum is preferentially scattered compared to the red end [Eq. (2.3)].

The effect of dissolved salts is to increase the molecular scattering of filtered seawater above that of pure water. The scattering coefficient v discussed below is about one third larger for "pure-seawater" than for distilled water. In the clear ocean water as much as 6% of the energy is "back scattered"; that is, if two light meters were placed back to back in the ocean, the one pointing downward would record an amount of "upwelled light" equal to 6% of that recorded by the upward-looking light meter. In more turbid water the ratio is decreased to about 2%.

The attenuation of light (β) [which includes both the effects of scattering (v) and absorption (ζ)] obeys Beers' law:

$$[-\beta(Z_2 - Z_1)]\,\Gamma_2 = \Gamma_1 \tag{12.15}$$

where Γ_2 and Γ_1 is the irradiance at depths Z_2 and Z_1, respectively, β is the attenuation coefficient, and $\beta = \zeta + v$. The irradiance is the flux of radiant energy incident at a point. The units of irradiance are energy per unit area per unit time, often langley per minute.

Absorption dominates the attenuation coefficient except in a narrow range of the visible spectrum. Figure 12.12 gives the attenuation coefficient for "pure seawater." For that part of the spectrum in the visible and near-visible range the scattering coefficient is also included. Scattering is an important contribution to the total attenuation coefficient in the 350- to 500-nm range, but not elsewhere. At all

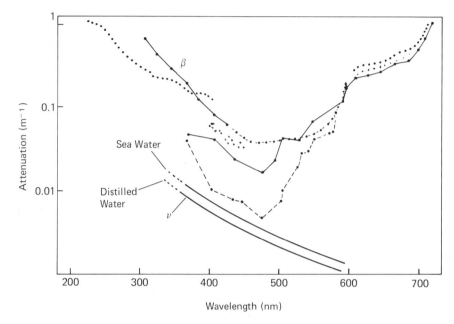

Figure 12.12. Scattering coefficient (v) is shown for distilled water and filtered seawater, (the two parallel lines near the bottom), as are data from various investigators for the attenuation in seawater (β). Losses due to absorption alone for a given wavelength would be the difference between β and v. (After Morel, A., 1974: "Optical Properties of Pure Water and Pure Sea Water," in *Optical Aspects of Oceanography*, eds. N. G. Jerlov and E. S. Nielson, Academic Press.)

other wavelengths the attenuation of light is primarily by absorption. Figure 12.13 is a dramatic representation of how narrow the transmission window for the electromagnetic spectrum is in water.

It is often convenient to classify water types by transmittance Π, the percentage of irradiance that passes a given distance. If $Z_2 - Z_1$ in Eq. (12.15) is 1 m, the transmittance per meter is simply

$$\Pi = \frac{\Gamma_2}{\Gamma_1} \tag{12.16}$$

Jerlov has attempted to classify the oceans and coastal waters in terms of their optical properties. The irradiance transmission for three ocean and nine coastal water types is shown in Figure 12.14. He has also estimated the total radiative energy that reaches a given depth (Table 3.1). Note that in even the clearest ocean water more

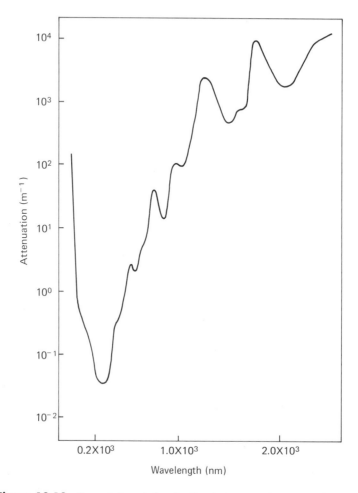

Figure 12.13. Transmission window for the electromagnetic spectrum is narrow and limited to the visible band. (After Morel, A., 1974: "Optical Properties of Pure Water and Pure Sea Water," in *Optical Aspects of Oceanography*, eds. N. G. Jerlov and E. S. Nielsen, Academic Press.)

than 50% of the radiation is absorbed in the first meter as a result of the immediate absorption of the infrared. Figure 12.14 indicates that the maximum transmission shifts toward higher wavelengths as the water becomes more turbid. In the clearest ocean water the maximum transmission is at about 465 nm (in the blue-green). In the most turbid waters the maximum transmission is about 575 nm. The reason is in part because particulate matter in the ocean appears to both scatter and absorb the shorter wavelengths more than the long, and because

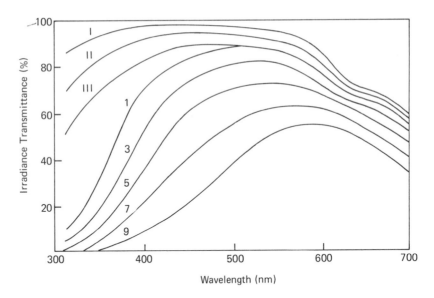

Figure 12.14. Radiation transmittance as a function of water type and wavelength. Roman numerals represent various oceanic water types. Arabic numerals are for coastal waters of increasing turbidity. (After Jerlov, N.G., 1968: Optical Oceanography, Elsevier Pub. Com., N.Y.)

dissolved organic matter in the ocean selectively absorbs the shorter wavelength. As a result, the wavelength of maximum transmittance shifts toward longer wavelengths in coastal regions where the total transmittance decreases.

A number of factors contribute to the color of the water, at least one of which has nothing to do with the ocean. The deep blue of the open ocean on a sunny day can turn grayish on an overcast day, because the blue of the ocean perceived by the eye while standing on the deck of a ship depends in part on the reflection from the ocean surface of the scattered sunlight, which is blue.

Under comparable atmospheric conditions, the color of the ocean is dependent upon its turbidity and the amount of dissolved organic matter. Oceanographers often refer to the deep blue ocean as being an "ocean desert," whereas a green ocean suggests high biological productivity. To the extent that this is true, it is primarily because of the presence of dissolved organic matter in the sea, which can vary from less than 1 mg/liter in the Sargasso Sea to as high as 3–4 mg/liter in rich coastal areas. The dissolved organic matter selectively absorbs the shorter wavelengths and causes a change in the ocean color.

   As might be expected, the eyes of creatures living in the ocean have adapted to that part of the spectrum with the greatest transmittance. It is probably not chance that the "visible spectrum" to us who have emerged from the ocean is that part of the radiation spectrum with the minimum extinction coefficients. Kampa and Boden have reported that there can even be adaptation within ocean regions. *Euphausia pacifica,* found in the open ocean, has maximum light sensitivity to 465 nm. A different variety of the same species found in Puget Sound is most sensitive at 495 nm.

appendix I

SELECTED DERIVATIONS

The derivations outlined here are neither complete nor rigorous. They are meant as a guide for those students who wish to know more about the basis for certain equations introduced and used in the text. A right-handed Cartesian coordinate system is used throughout with x, y, z, and $\mathbf{i}$, $\mathbf{j}$, $\mathbf{k}$ and u, v, w the space coordinates, unit vectors, and velocity components, respectively. The z coordinate is parallel to the gravity vector and is positive in the "upward" direction. Boldfaced symbols are used for vectors. Subscripts are often used to signify partial differentials. For example,

$$u_t \equiv \frac{\partial u}{\partial t}, \qquad \phi_{xz} \equiv \frac{\partial^2 \phi}{\partial x\, \partial z}$$

Although most derivations are done in terms of partial differential equations, the vector equivalent solution is usually given in addition.

Conservation Equations: Equations of Continuity

Consider a small cube of sizes Δx, Δy, Δz within a larger volume of fluid (Figure I.1). The density of fluid in the cubic element is ρ. The mass is therefore density times the volume, $m = \rho\, \Delta x\, \Delta y\, \Delta z$. We may assume that there is flow in and out of all sides of the cube, but consider first the flow in the x direction. The rate at which mass is entering the cube is $\rho_1 u_1\, \Delta y\, \Delta z$, and the rate at which it is leaving is $\rho_2 u_2\, \Delta y\, \Delta z$. The rate of change is

$$\frac{\partial m}{\partial t} = \frac{\partial}{\partial t}(\rho\, \Delta x\, \Delta y\, \Delta z) = \rho_1 u_1\, \Delta y\, \Delta z - \rho_2 u_2\, \Delta y\, \Delta z$$

$$\frac{\partial \rho}{\partial t} = \frac{\rho_1 u_1}{\Delta x} - \frac{\rho_2 u_2}{\Delta x}$$

We may assume that both density and velocity change continuously across the small cube and that

$$u_2 = u_1 + \Delta u$$

$$\rho_2 = \rho_1 + \Delta \rho$$

$$u_2 = u + \frac{\Delta u}{2}, \qquad u_1 = u - \frac{\Delta u}{2}$$

$$\rho_2 = \rho + \frac{\Delta \rho}{2}, \qquad \rho_1 = \rho - \frac{\Delta \rho}{2}$$

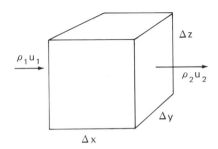

Figure I.1.

Substituting

$$\frac{\partial \rho}{\partial t} \Delta x = \left(u - \frac{\Delta u}{2} \right) \left(\rho - \frac{\Delta \rho}{2} \right) - \left(u + \frac{\Delta u}{2} \right) \left(\rho + \frac{\Delta \rho}{2} \right)$$

$$\frac{\partial \rho}{\partial t} = -\rho \frac{\Delta u}{\Delta x} - u \frac{\Delta \rho}{\Delta x}$$

As the cubic element is reduced to a differential, the terms on the right become

$$\frac{\partial \rho}{\partial t} = -\rho \frac{\partial u}{\partial x} - u \frac{\partial \rho}{\partial x} = -\frac{\partial}{\partial x} (\rho u)$$

Although the negative sign falls out naturally in the example used, the reader can easily convince himself that the relationship is general by working through the example with u_1 and/or ρ_1 larger than u_2 and/or ρ_2, or by changing the direction of u_1 and/or u_2. Similarly, the derivation can be repeated for the other two directions, which leads to the general form

$$\frac{\partial \rho}{\partial t} = -\frac{\partial}{\partial x} (\rho u) - \frac{\partial}{\partial y} (\rho v) - \frac{\partial}{\partial z} (\rho w) \tag{I.1}$$

In the case of a fluid of constant density

$$\frac{\partial \rho}{\partial t} = \frac{\partial \rho}{\partial x} = \frac{\partial \rho}{\partial y} = \frac{\partial \rho}{\partial z} = 0$$

In this case the equation of continuity is

$$\frac{\partial u}{\partial x} + \frac{\partial v}{\partial y} + \frac{\partial w}{\partial z} = 0 \tag{I.2}$$

or in vector notation
$$\nabla \cdot \mathbf{V} = 0 \tag{I.3}$$

where ∇ defines the operator

$$\nabla = \mathbf{i}\frac{\partial}{\partial x} + \mathbf{j}\frac{\partial}{\partial y} + \mathbf{k}\frac{\partial}{\partial z}$$

and

$$\mathbf{V} = \mathbf{i}u + \mathbf{j}v + \mathbf{k}w$$

Equations I.2 and I.3 are the continuity equations or the conservation of mass equations for an incompressible, homogeneous fluid. Although the assumption of constant density in the oceans is not exact, it is sufficiently close for nearly all hydrodynamic calculations requiring the equation of continuity.

The derivation for the conservation of salt (and other conservative properties) is similar to that for the continuity equation (the conservation of mass). Salt is usually measured in terms of salinity,

$$S = \frac{\text{grams of salt}}{\text{kilogram of water}} = \frac{m_s}{m_w} \times 10^3$$

where m_s is the mass of salt and m_w is the mass of seawater. The product $S\rho$ is in units of mass per unit volume, which is required for the derivation.

By analogy with the continuity equation, the flow of salt into the unit cube in the x direction is $S_1\rho_1 u_1 \, \Delta y \, \Delta z$, and the flow out is $S_2\rho_2 u_2 \, \Delta y \, \Delta z$. Using identical arguments, it can be shown that the change in the mass of salt within the cube resulting from these flows is

$$\frac{\partial}{\partial t}\,(S\rho)\,\Delta x\,\Delta y\,\Delta z = S_1\rho_1 u_1\,\Delta y\,\Delta z - S_2\rho_2 u_2\,\Delta y\,\Delta z$$

and

$$\frac{\partial}{\partial t}\,(S\rho) = -\frac{\partial}{\partial x}(S\rho u)$$

Carrying out the same argument in the other two dimensions yields

$$\frac{\partial}{\partial t}(S\rho) = -\frac{\partial}{\partial x}(S\rho u) - \frac{\partial}{\partial y}(S\rho v) - \frac{\partial}{\partial z}(S\rho w)$$

Expanding the above equation gives

$$S\frac{\partial \rho}{\partial t} + \rho\frac{\partial S}{\partial t} = -\rho\left(u\frac{\partial S}{\partial x} + v\frac{\partial S}{\partial y} + w\frac{\partial S}{\partial z}\right) - S\rho\left(\frac{\partial u}{\partial x} + \frac{\partial v}{\partial y} + \frac{\partial w}{\partial z}\right)$$
$$- S\left(u\frac{\partial \rho}{\partial x} + v\frac{\partial \rho}{\partial y} + w\frac{\partial \rho}{\partial z}\right)$$

Making the same assumption about constant density as in the previous section gives the final form:

$$\frac{\partial S}{\partial t} = -u\frac{\partial S}{\partial x} - v\frac{\partial S}{\partial y} - w\frac{\partial S}{\partial z} \tag{I.4}$$

$$\frac{\partial S}{\partial t} = -(\mathbf{V}\cdot\nabla)S \tag{I.5}$$

Acceleration

In a fluid, velocity is not only a function of time, but also of space:

$$u = f(x, y, z, t)$$

By the chain rule of differentiation,

$$\frac{du}{dt} = \frac{\partial u}{\partial t} + \frac{\partial u}{\partial x}\frac{dx}{dt} + \frac{\partial u}{\partial y}\frac{dy}{dt} + \frac{\partial u}{\partial z}\frac{dz}{dt}$$

$$= \frac{\partial u}{\partial t} + u\frac{\partial u}{\partial x} + v\frac{\partial u}{\partial y} + w\frac{\partial u}{\partial z}$$

For emphasis, the total differential is often written Du/Dt:

$$\frac{Du}{Dt} \equiv \frac{du}{dt} = u_t + uu_x + vu_y + wu_z \tag{I.6}$$

where the subscript refers to a partial differential. Similarly,

$$\frac{dv}{dt} = v_t + uv_x + vv_y + wv_z$$

$$\left.\begin{array}{c} \\ \\ \\ \end{array}\right\} \quad \text{(I.6)}$$

$$\frac{dw}{dt} = w_t + uw_x + vw_y + ww_z$$

In vector notation,

$$\frac{du}{dt} = u_t + (\mathbf{V}\cdot\mathbf{\nabla})u$$

$$\frac{dv}{dt} = v_t + (\mathbf{V}\cdot\mathbf{\nabla})v \qquad \left.\begin{array}{c} \\ \\ \\ \\ \end{array}\right\} \quad \text{(I.7)}$$

$$\frac{dw}{dt} = w_t + (\mathbf{V}\cdot\mathbf{\nabla})w$$

or in general

$$\frac{d\mathbf{V}}{dt} = \frac{\partial\mathbf{V}}{\partial t} + (\mathbf{V}\cdot\mathbf{\nabla})\mathbf{V} \qquad\qquad \text{(I.8)}$$

Note that d/dt or D/Dt is the particle acceleration, and $\partial/\partial t$ is the local acceleration. In the previously derived conservation equation we used the local change of mass $\partial m/\partial t$. Why?

Pressure Gradient

Consider a cube of fluid of density ρ with sides Δx, Δy, Δz, and let this element of fluid be in a channel where the pressure increases from left to right (i.e., $p_2 > p_1$; Figure I.2). Remembering that a pressure force is pressure times the cross-sectional area, with the force vector acting normal to the cross section, the force on the two sides of the cube would be

$$F_1 = p_1\Delta y\ \Delta z, \qquad F_2 = p_2\Delta y\ \Delta z$$

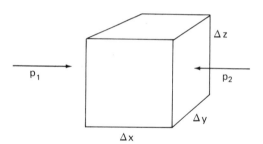

Figure I.2.

Let the pressure p_2 be slightly larger than p_1:

$$p_2 = p_1 + \Delta p$$

The mass of fluid element is simply the density times the volume

$$m = \rho \, \Delta x \, \Delta y \, \Delta z$$

Equating the acceleration of the cubic mass to the pressure forces,

$$\frac{du}{dt} (\rho \, \Delta x \, \Delta y \, \Delta z) = F_1 - F_2 = p_1 \, \Delta y \, \Delta z - (p_1 + \Delta p) \, \Delta y \, \Delta z$$

$$\frac{du}{dt} = -\frac{1}{\rho} \frac{\Delta p}{\Delta x}$$

By letting the cube of fluid become very small, one approaches the differential form

$$\frac{du}{dt} = -\frac{1}{\rho} \frac{\partial p}{\partial x} = -\frac{1}{\rho} p_x$$

Note that F_2 was given a negative sign because the force was directed in the $-x$ direction. The meaning of the negative sign in the final equation is simply that the particle is accelerated from high pressure toward low pressure. A similar derivation can be done in the other two directions. Combined, these give

$$\frac{du}{dt} = -\frac{1}{\rho} p_x$$

$$\frac{dv}{dt} = -\frac{1}{\rho} p_y \qquad (I.9)$$

$$\frac{dw}{dt} = -\frac{1}{\rho} p_z$$

In vector notation these become

$$\frac{d\mathbf{V}}{dt} = -\frac{1}{\rho} \boldsymbol{\nabla} p \qquad (I.10)$$

Coriolis Acceleration

The derivation of the Coriolis term can best be done with vector algebra. Take the center of the earth as the origin of the coordinate system. A point on the surface of the earth is given by

$$\mathbf{R} = \mathbf{i}x + \mathbf{j}y + \mathbf{k}z \qquad (I.11)$$

Note that $\mathbf{i}$, $\mathbf{j}$, and $\mathbf{k}$ are still east, north, and up with respect to the print on the surface of the earth (Figure I.3).

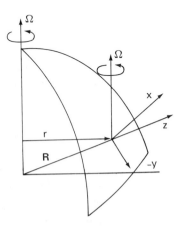

Figure I.3.

· Taking the derivative of **R** with respect to time yields two sets of terms:

$$\frac{d\mathbf{R}}{dt} = (\mathbf{i}x_t + \mathbf{j}y_t + \mathbf{k}z_t) + (x\mathbf{i}_t + y\mathbf{j}_t + z\mathbf{k}_t) \tag{I.12}$$

The first term on the right is the movement of that point on the surface of the earth relative to a fixed coordinate system on the earth. It is the term one usually refers to as velocity (**V**), and it is the movement an observer on earth would measure when he ignores (as we always do) the rotation of the earth and its movement through space. We shall call this term the *relative velocity* $\dot{\mathbf{R}}$:

$$\dot{\mathbf{R}} = \mathbf{i}\dot{x} + \mathbf{j}\dot{y} + \mathbf{k}\dot{z}$$

where the dot means the derivative with respect to time (i.e., $\dot{x} \equiv x_t$).

The second term is the movement of the coordinate system; it is the movement of a fixed point on the earth relative to the origin. The latter is simply the cross product of the vector radius and the angular velocity of the earth:

$$x\mathbf{i}_t + y\mathbf{j}_t + z\mathbf{k}_t = \mathbf{\Omega} \times \mathbf{R}$$

Thus the movement of the point on the surface of the earth, relative to a coordinate system whose origin is the center of the earth, is of two kinds: movement relative to a fixed coordinate system on the *surface* of the earth and the rotation of that fixed coordinate system.

$$\frac{d\mathbf{R}}{dt} = \dot{\mathbf{R}} + \mathbf{\Omega} \times \mathbf{R} \tag{I.13}$$

The next step is to take the second derivative of **R** with time. If the reader wishes, he can take the derivative of Eq. (I.12) and then proceed to sort out the resulting 12 terms. A simpler approach is to note that Eq. (I.13) is quite general and defines an operator

$$\frac{d}{dt} = \dot{} + \mathbf{\Omega} \times \tag{I.14}$$

Then

$$\frac{d}{dt}\left(\frac{d\mathbf{R}}{dt}\right) = (\dot{\ } + \mathbf{\Omega} \times) (\dot{\mathbf{R}} + \mathbf{\Omega} \times \mathbf{R})$$

$$\frac{d^2\mathbf{R}}{dt^2} = \ddot{\mathbf{R}} + \mathbf{\Omega} \times \dot{\mathbf{R}} + \dot{\mathbf{\Omega}} \times \mathbf{R} + \mathbf{\Omega} \times \dot{\mathbf{R}} + \mathbf{\Omega} \times \mathbf{\Omega} \times \mathbf{R}$$

Since we assume that the angular rotation of the earth is constant, the third term on the right is zero.

$$\frac{d^2\mathbf{R}}{dt^2} = \ddot{\mathbf{R}} + 2\mathbf{\Omega} \times \dot{\mathbf{R}} + \mathbf{\Omega} \times \mathbf{\Omega} \times \mathbf{R} \tag{I.15}$$

The term on the left side is the acceleration of a point on the surface of the earth measured with respect to a coordinate system whose origin is the center of the earth. The first term on the right is the acceleration relative to the fixed coordinate system on the surface of the earth. It is the acceleration an observer on the surface of the earth would measure when he ignores (as we always do) the rotation of the earth and its movement through space. The second term represents the Coriolis acceleration. The third is related to what is usually considered a part of the earth's gravitational field. The Coriolis term relates the acceleration measured with respect to a coordinate system whose origin is the center of the earth to the acceleration measured with respect to a coordinate system at a fixed point on a rotating earth.

The final step is to translate the Coriolis terms to the fixed coordinate system on the surface of the earth. Using the usual notation of $\mathbf{i}$, x is east, $\mathbf{j}$, y is north, and $\mathbf{k}$, w is up,

$$\mathbf{R} = \mathbf{k}R$$

$$\dot{\mathbf{R}} = \mathbf{i}u + \mathbf{j}v + \mathbf{k}w$$

$$\mathbf{\Omega} = \mathbf{j}\Omega \cos \phi + \mathbf{k}\Omega \sin \phi$$

where ϕ is latitude,

$$2\mathbf{\Omega} \times \dot{\mathbf{R}} = 2 \begin{vmatrix} \mathbf{i} & \mathbf{j} & \mathbf{k} \\ 0 & \Omega \cos \phi & \Omega \sin \phi \\ u & v & w \end{vmatrix}$$

$$= 2\mathbf{i}(w\Omega \cos \phi - v\Omega \sin \phi) + 2\mathbf{j}(0 + u\Omega \sin \phi)$$
$$+ 2\mathbf{k}(\mathrm{o} - u\Omega \cos \phi) \qquad (\mathrm{I}.16)$$

$$= \mathbf{j}(2u\Omega \sin \phi) - \mathbf{i}(2v\Omega \sin \phi) + \mathbf{i}(2w\Omega \cos \phi)$$
$$- \mathbf{k}(2u\Omega \cos \phi)$$

The first two terms are those usually referred to as the *Coriolis acceleration* in oceanography. The third term is usually ignored, because the average vertical velocities in the ocean are considered to be one to several orders of magnitude less than the horizontal velocities. Of course, this term cannot be ignored in problems of ballistics, where the vertical velocity may be of the same order or larger than the horizontal terms. The fourth term is in the direction of gravity. It is the Eötvös correction (see discussion of gravity in Chapter 5).

In a similar manner, it can be shown that

$$\mathbf{\Omega} \times (\mathbf{\Omega} \times \mathbf{R}) = \mathbf{j}\Omega^2 r \sin \phi - \mathbf{k}\Omega^2 r \cos \phi \qquad (\mathrm{I}.17)$$

where $r = R \cos \phi$ and is the distance of a point on the earth from the earth's axis. Note that these terms are a function of position only and are independent of velocity. They are important in determining the "gravitational field" of the earth (see discussion of gravity in Chapter 5).

In the absence of any external forces, the left side of Eq. (I.15) is zero. Ignoring the "gravitational" term, Eq. (I.17), and remembering that $\dot{\mathbf{R}} \equiv \mathbf{V}$, Eq. (I.15) reduces to

$$\frac{d\mathbf{V}}{dt} = -2\mathbf{\Omega} \times \mathbf{V} \qquad (\mathrm{I}.18)$$

$$\frac{du}{dt} = +fv - w2\Omega \cos \phi$$
$$\qquad (\mathrm{I}.19)$$
$$\frac{dv}{dt} = -fu$$

where $f = 2\Omega \sin \phi$. The terms on the right side of Eqs. (I.18. and (I.19) are the *Coriolis force* terms. As noted earlier, the vertical velocity in the ocean is generally many times smaller than the horizontal velocity, and the term $w2\Omega \cos \phi$ is generally ignored.

Equation of Motion

By combining the various force terms of equations, we can now write the equations of motion for a fluid. In the horizontal plane they are

$$u_t + uu_x + vu_y + wu_z = -\frac{1}{\rho}p_x + fv + \mathcal{X}$$

$$v_t + uv_x + vv_y + wv_z = -\frac{1}{\rho}p_y - fu + \mathcal{Y}$$

(I.20)

where $\mathcal{X}$ and $\mathcal{Y}$ are undefined frictional forces. In vector notation,

$$\frac{\partial \mathbf{V}}{\partial t} + (\mathbf{V} \cdot \nabla)\mathbf{V} = -\frac{1}{\rho}\nabla p - 2\Omega \times \mathbf{V} + \mathbf{g} + \boldsymbol{\phi}$$

(I.21)

where $\boldsymbol{\phi}$ is an undefined vectorial friction term and $\mathbf{g}$ is gravity.

Friction Shearing Stresses (*Molecular and Eddy Viscosity*)

A viscous fluid is subject to shearing stresses. For example, if a wind blows parallel to a fluid surface, the surface of the fluid begins to move in the direction of the wind, as momentum is transferred to the interior of the fluid by random molecular motion. This is the process of molecular viscosity.

$$\tau_{xz} = \mu \frac{\partial u}{\partial z}$$

where τ_{xz} is the shearing stress tensor. The first subscript refers to the direction of the stress; the second subscript refers to the plane in which the stress acts. The subscript z is normal to the xy plane in which the stress acts. μ is the molecular viscosity. Shearing stress has the same units as pressure. The force applied on the surface of the fluid is stress times surface area. There are three elements of stress tensor acting in each of three directions. In the x direction they are

$$
\left.
\begin{aligned}
\tau_{xz} &= \mu \frac{\partial u}{\partial z} \\[2ex]
\tau_{xy} &= \mu \frac{\partial u}{\partial y} \\[2ex]
\tau_{xx} &= \mu \frac{\partial u}{\partial x}
\end{aligned}
\right\}
\tag{I.22}
$$

Consider the force in the x direction that these stresses apply to a fluid element (Figure I.4). The force applied in the cube is the difference between the shearing forces in the upper and lower faces.

In terms of the acceleration of the mass of the cube,

$$
m \frac{du}{dt} = \rho \, \Delta x \, \Delta y \, \Delta z \, \frac{du}{dt} = \tau_{xz_2} (\Delta x \, \Delta y) - \tau_{xz_1} (\Delta x \, \Delta y)
$$

$$
\frac{du}{dt} = \frac{1}{\rho} \frac{\tau_{xz_2} - \tau_{xz_1}}{\Delta z} = \frac{1}{\rho} \frac{\Delta \tau_{xz}}{\Delta z}
$$

In the limit as the cube becomes small, this reduces to

$$
\frac{du}{dt} = \frac{1}{\rho} \frac{\partial \tau_{xz}}{\partial z} = \frac{1}{\rho} \frac{\partial}{\partial z} \left(\mu \frac{\partial u}{\partial z} \right)
$$

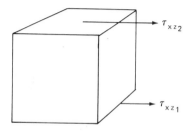

Figure I.4.

$$= \frac{1}{\rho} \mu \frac{\partial^2 u}{\partial z^2}$$

By a similar argument, the other stress terms can be found, and

$$\frac{du}{dt} = \frac{1}{\rho} \mu \frac{\partial^2 u}{\partial z^2} + \frac{1}{\rho} \mu \frac{\partial^2 u}{\partial y^2} + \frac{1}{\rho} \mu \frac{\partial^2 u}{\partial x^2}$$

$$\frac{dv}{dt} = \frac{1}{\rho} \mu \frac{\partial^2 v}{\partial z^2} + \frac{1}{\rho} \mu \frac{\partial^2 v}{\partial y^2} + \frac{1}{\rho} \mu \frac{\partial^2 v}{\partial x^2} \qquad (\text{I.23})$$

$$\frac{dw}{dt} = \frac{1}{\rho} \mu \frac{\partial^2 w}{\partial z^2} + \frac{1}{\rho} \mu \frac{\partial^2 w}{\partial y^2} + \frac{1}{\rho} \mu \frac{\partial^2 w}{\partial x^2}$$

$$\frac{d\mathbf{V}}{dt} = \frac{\mu}{\rho} \nabla^2 \mathbf{V} \qquad (\text{I.24})$$

The molecular viscosity of water is of the order of 10^{-2} g/cm/s. The highest values of time averaged velocity shear observed in the ocean are of the order of 5×10^{-2}/s. Thus it is easy to demonstrate that molecular viscosity is not an important term in the equation of motion in oceanography.

The ocean is fully turbulent. Swirls and eddies of dimensions from tens of centimeters to tens of kilometers are thought to occur and move at random within the ocean. One can imagine these eddies as transferring momentum in a manner analogous to that performed by random molecular motion in the case of molecular viscosity. It is then possible to define an eddy viscosity coefficient or Austausch coefficient (A) analogous to the molecular viscosity coefficient. By analogous arguments, one can derive the eddy viscosity version of Eq. (I.23).

$$\frac{du}{dt} = \frac{1}{\rho} \frac{\partial}{\partial z} \left(A_z \frac{\partial u}{\partial z} \right) + \frac{1}{\rho} \frac{\partial}{\partial y} \left(A_y \frac{\partial u}{\partial y} \right) + \frac{1}{\rho} \frac{\partial}{\partial x} \left(A_x \frac{\partial u}{\partial x} \right)$$

$$\frac{dv}{dt} = \frac{1}{\rho} \frac{\partial}{\partial z} \left(A_z \frac{\partial v}{\partial z} \right) + \frac{1}{\rho} \frac{\partial}{\partial y} \left(A_y \frac{\partial v}{\partial y} \right) + \frac{1}{\rho} \frac{\partial}{\partial x} \left(A_x \frac{\partial v}{\partial x} \right) \qquad (\text{I.25})$$

$$\frac{dw}{dt} = \frac{1}{\rho} \frac{\partial}{\partial z} \left(A_z \frac{\partial w}{\partial z} \right) + \frac{1}{\rho} \frac{\partial}{\partial y} \left(A_y \frac{\partial w}{\partial y} \right) + \frac{1}{\rho} \frac{\partial}{\partial x} \left(A_x \frac{\partial w}{\partial x} \right)$$

In Eq. (I.25) it is not assumed that A is constant. However, such an assumption is made in most cases where these terms are used. When constant, the eddy coefficient would be taken outside the second differential. As written, Eq. (I.25) allows for different values of A in the x, y, z direction. The usual assumption is that

$$A_x = A_y = A_h$$

where the subscript refers to horizontal eddy viscosity. Because the ocean is vertically stratified, work is required to move turbulent eddies vertically. Thus it is supposed that $A_h \gg A_z$. In terms of Eq. (I.20), we can write

$$
\left.
\begin{aligned}
u_t + uu_x + vu_y + wu_z &= -\frac{1}{\rho} p_x + fv + \frac{A_h}{\rho}(u_{xx} + u_{yy}) \\
&\quad + \frac{A_z}{\rho} u_{zz} \\
v_t + uv_x + vv_y + wv_z &= -\frac{1}{\rho} p_y - fu + \frac{A_h}{\rho}(v_{xx} + v_{yy}) \\
&\quad + \frac{A_z}{\rho} v_{zz}
\end{aligned}
\right\} \quad \text{(I.26)}
$$

Note that A_z is the vertical component of the Austausch coefficient and *not* $\partial A/\partial z$.

Reynolds Stress Terms

The key to deriving the Reynolds stress terms is understanding the averaging process. Any scaler velocity component can be averaged over time:

$$\bar{u} = \frac{1}{T} \int_0^T u \, dt \tag{I.27}$$

where u is the instantaneous velocity component and $\bar{u}$ the average value of that component as observed over time, T. The instantaneous velocity can be written in terms of its average value and a fluctuating component, u':

$$u = \bar{u} + u'$$

It follows that the average value of u', averaged over the same period, must be zero:

$$\overline{u'} = \frac{1}{T} \int_0^T u' \, dt = 0 \qquad (\text{I.28})$$

Consider next the averaging of the product of two separate components u and v:

$$uv = (\overline{u} + u')(\overline{v} + v') = \overline{u}\overline{v} + \overline{u}v' + \overline{v}u' + u'v'$$

Now take the average value of each of the four terms:

$$\overline{uv} = \frac{1}{T} \int_0^T uv \, dt = \frac{1}{T} \int_0^T \overline{u}\overline{v} \, dt + \frac{1}{T} \int_0^T \overline{u}v' \, dt + \frac{1}{T} \int_0^T \overline{v}u' \, dt$$

$$+ \frac{1}{T} \int_0^T u'v' \, dt$$

Thus

$$\overline{uv} = \overline{\overline{u}\overline{v}} + \overline{\overline{u}v'} + \overline{\overline{v}u'} + \overline{u'v'}$$

Since $\overline{u}, \overline{v}$ are constant over the interval T,

$$\overline{uv} = \overline{u}\overline{v} + \overline{u}\overline{v'} + \overline{v}\overline{u'} + \overline{u'v'}$$

But since $\overline{v'} = \overline{u'} = 0$,

$$\overline{uv} = \overline{u}\overline{v} + \overline{u'v'} \qquad (\text{I.29})$$

Note that, although the average value of the fluctuating velocity component u' and v' is zero, the average value of the product of these fluctuating components *need not* be zero.

To apply this averaging process to the equation of motion, it is first necessary to manipulate it a bit. Ignoring friction, the x component of the equation of motion, Eq. (I.20), can be written

$$\rho u_t + \rho u u_x + \rho v u_y + \rho w u_z = -p_x + \rho f v \qquad (\text{I.30})$$

The equation of continuity, Eq. (I.1), is

$$\rho_t + (\rho u)_x + (\rho v)_y + (\rho w)_z = 0 \qquad\qquad \text{(I.1)}$$

Multiply the continuity equation by u:

$$u\rho_t + u(\rho u)_x + u(\rho v)_y + u(\rho w)_z = 0$$

Add the above equation to the x component of the equation of motion:

$$[\rho u_t + u\rho_t] + [u(\rho u)_x + \rho u u_x] + [u(\rho v)_y + \rho v u_y] + [u(\rho w)_z + \rho w u_z] = -p_x + \rho f v$$

which can be written

$$(u\rho)_t + (\rho u u)_x + (\rho u v)_y + (\rho u w)_z = -p_x + \rho f v$$

Next substitute $\bar{u} + u'$, and so on, for the instantaneous velocity components and then average:

$$\overline{[(\bar{u} + u')\rho]_t} + \overline{[\rho(\bar{u} + u')(\bar{u} + u')]_x} + \overline{[\rho(\bar{u} + u')(\bar{v} + v')]_y}$$

$$+ \overline{[\rho(\bar{u} + u')(\bar{w} + w')]_z} = -\bar{p}_x + \overline{\rho f(\bar{v} + v')}$$

This reduces to

$$(\bar{u}\rho)_t + [\rho(\overline{\bar{u}\bar{u}} + \overline{u'u'})]_x + [\rho(\overline{\bar{u}\bar{v}} + \overline{u'v'})]_y + [\rho(\overline{\bar{u}\bar{w}} + \overline{w'u'})]_z = -p_x + \rho f \bar{v}$$

The averaging bars have been dropped over terms assumed constant over the time interval. The time derivative is assumed to be for intervals long compared to the time over which $\bar{u}$ is averaged in Eq. (I.27). Rearranging terms,

$$(\bar{u}\rho)_t + (\overline{\rho\bar{u}\bar{u}})_x + (\overline{\rho\bar{u}\bar{v}})_y + (\overline{\rho\bar{u}\bar{w}})_z = -p_x + \rho f \bar{v} - \overline{(\rho u'u')}_x$$

$$- \overline{(\rho u'v')}_y - \overline{(\rho u'w')}_z$$

The above equation is identical in form with Eq. (I.30) except that average rather than instantaneous velocity components are used, and there are three additional terms representing averages of cross products of the fluctuating components.

By reversing the manipulation done to arrive at Eq. (I.30), one can write the x component of the equation of motion as

$$\bar{u}_t + \overline{\bar{u}\bar{u}}_x + \overline{\bar{v}\bar{u}}_y + \overline{\bar{w}\bar{u}}_z = -\frac{1}{\rho}\bar{p}_x + \bar{f}\bar{v} - \overline{(u'u')}_x - \overline{(u'v')}_y$$
$$- \overline{(u'w')}_z$$

The last three terms are called the *Reynolds stress terms*. A similar derivation can be made for the y component. Where the averaging processes implicit with Reynolds stress consideration is understood, the averaging bars can be dropped, and we can write

$$\left.\begin{aligned}
u_t + uu_x + vu_y + wu_z &= -\frac{1}{\rho}p_x + fv - \overline{(u'u')}_x \\
&\quad - \overline{(u'v')}_y - \overline{(u'w')}_z \\
v_t + uv_x + vv_y + wv_z &= -\frac{1}{\rho}p_y - fu - \overline{(v'u')}_x \\
&\quad - \overline{(v'v')}_y - \overline{(v'w')}_z
\end{aligned}\right\} \quad \text{(I.31)}$$

Prandtl's mixing-length theory then relates the Reynolds stresses to the shears in the mean velocities by relations of the form

$$\left.\begin{aligned}
\overline{u'u'} = \frac{-A_x}{\rho}\frac{\partial\bar{u}}{\partial x}, \quad \overline{u'v'} = \frac{-A_y}{\rho}\frac{\partial\bar{u}}{\partial y}, \quad \overline{u'w'} = \frac{-A_z}{\rho}\frac{\partial\bar{u}}{\partial z} \\
\overline{v'u'} = \frac{-A_x}{\rho}\frac{\partial\bar{v}}{\partial x}, \quad \overline{v'v'} = \frac{-A_y}{\rho}\frac{\partial\bar{v}}{\partial y}, \quad \overline{v'w'} = \frac{-A_z}{\rho}\frac{\partial\bar{v}}{\partial z}
\end{aligned}\right\} \text{(I.32)}$$

where the A's are the Austausch coefficients.

Diffusion Terms in the Conservation Equation

Consider the changes in salinity within a unit cube caused by diffusion. The salinity gradient in the x direction in the water surrounding the cube is shown in Figure I.5. Salt is being diffused from high salinity toward low salinity. The change in salinity within the cube is a function of the difference in the diffusion of salt across the two sides of the cube where j is the coefficient of molecular diffusion.

$$-\frac{\partial}{\partial t}(S\rho)\,\Delta x\,\Delta y\,\Delta z = F_2\,\Delta y\,\Delta z - F_1\,\Delta y\,\Delta z =$$

$$\left[-\left(j\,\frac{\partial S}{\partial x}\right)_2 + \left(j\,\frac{\partial S}{\partial x}\right)_1\right]\Delta y\,\Delta z$$

$$\rho\,\frac{\partial S}{\partial t} = +\frac{j}{\Delta x}\left[\left(\frac{\partial S}{\partial x}\right)_2 - \left(\frac{\partial S}{\partial x}\right)_1\right]$$

Letting the cube become very small, we can write the differential

$$\frac{\partial S}{\partial t} = +\frac{j}{\rho}\frac{\partial^2 S}{\partial x^2}$$

Assuming that there is a salinity gradient in all three dimensions, we can show that the change in salinity within the cube caused by molecular diffusion is

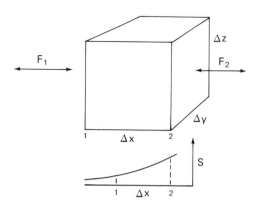

Figure I.5.

287

$$\frac{\partial S}{\partial t} = + \frac{j}{\rho} \left(\frac{\partial^2 S}{\partial x^2} + \frac{\partial^2 S}{\partial y^2} + \frac{\partial^2 S}{\partial z^2} \right) \tag{I.33}$$

$$\frac{\partial S}{\partial t} = + \frac{j}{\rho} \nabla^2 S \tag{I.34}$$

Similar equations can be written for the diffusion of heat.

On the scale of meters or larger, molecular processes do not play an important role in changing the distribution of properties in the ocean. (Double-diffusion processes such as salt fingers may be an exception, see Chapter 9). However, diffusion can occur through turbulence, and, as in the case of eddy viscosity, one can argue that the movement of eddies and swirls plays an analogous role to that of the random motion of molecules. Thus one can write the flux equations with eddy diffusion coefficients replacing the molecular diffusion coefficients. Since the physical arguments are the same, the form of the initial equations is identical. The derivation is the same, and one arrives at

$$\frac{\partial S}{\partial t} = + \frac{\partial}{\partial x} \left(\frac{A_x}{\rho} \frac{\partial S}{\partial x} \right) + \frac{\partial}{\partial y} \left(\frac{A_y}{\rho} \frac{\partial S}{\partial y} \right) + \frac{\partial}{\partial z} \left(\frac{A_z}{\rho} \frac{\partial S}{\partial z} \right) \tag{I.35}$$

As written, the eddy coefficients are not assumed to be independent of space. Furthermore, it is not assumed that they are identical. Because the ocean is vertically stratified, turbulent processes are depressed in the vertical as compared to the horizontal. Thus A_z should be smaller than A_x or A_y; the latter two are generally considered to be equal, and the notation A_h for horizontal diffusion is used. Assuming that the diffusion coefficients are constant, Eq. (I.35) can be written

$$\frac{\partial S}{\partial t} = + \frac{A_h}{\rho} \left(\frac{\partial^2 S}{\partial x^2} + \frac{\partial^2 S}{\partial y^2} \right) + \frac{A_z}{\rho} \frac{\partial^2 S}{\partial z^2} \tag{I.36}$$

Considering both the diffusion as well as the advection of material requires rewriting the conservation equations. For example, Eqs. (I.4) and (I.5) become

$$-\frac{\partial S}{\partial t} = u\,\frac{\partial S}{\partial x} + v\,\frac{\partial S}{\partial y} + w\,\frac{\partial S}{\partial z} - \frac{A_h}{\rho}\left(\frac{\partial^2 S}{\partial x^2} + \frac{\partial^2 S}{\partial y^2}\right)$$
$$- \frac{A_z}{\rho}\,\frac{\partial^2 S}{\partial z^2} \tag{I.37}$$

$$-\frac{\partial S}{\partial t} = (\mathbf{V}\cdot\nabla)S - \frac{\mathbf{A}}{\rho}\cdot\nabla^2 S \tag{I.38}$$

Diffusion can also be written in terms of turbulent perturbations about the mean. The conservation equation for an incompressible fluid of constant density, Eqs. (I.2) and (I.4), can be written

$$-\frac{\partial S}{\partial t} = \frac{\partial}{\partial x}(Su) + \frac{\partial}{\partial y}(Sv) + \frac{\partial}{\partial z}(Sw) \tag{I.39}$$

Using the ideas and notation developed for the derivation of Reynolds stresses, Eq. (I.39) becomes

$$-\frac{\partial S}{\partial t} = +\frac{\partial}{\partial x}(\overline{uS}) + \frac{\partial}{\partial y}(\overline{vS}) + \frac{\partial}{\partial z}(\overline{wS}) + \frac{\partial}{\partial x}\overline{(S'u')}$$
$$+ \frac{\partial}{\partial y}\overline{(S'v')} + \frac{\partial}{\partial z}\overline{(S'w')} \tag{I.40}$$

There is a formal equivalence between the eddy diffusion and turbulent transfer terms:

$$\left.\begin{aligned}
-\overline{(S'u')} &= \frac{A_h}{\rho}\,\frac{\partial S}{\partial x}\\[6pt]
-\overline{(S'v')} &= \frac{A_h}{\rho}\,\frac{\partial S}{\partial x}\\[6pt]
-\overline{(S'w')} &= \frac{A_z}{\rho}\,\frac{\partial S}{\partial z}
\end{aligned}\right\} \tag{I.41}$$

Vorticity

Vorticity can be defined as the cross product of velocity:

$$\nabla\times\mathbf{V} = \mathbf{i}\,(w_y - v_z) + \mathbf{j}\,(u_z - w_x) + \mathbf{k}\,(v_x - u_y) \tag{I.42}$$

When $\nabla \times V = 0$, the motion is said to be irrotational. The two horizontal components of the equation of motion, Eq. (I.20), are often written in the form of a single vorticity equation in terms of the *horizontal vorticity*, which is defined as $\xi = v_x - u_y$. We begin with the horizontal components of the equation of motions.

$$v_t + uv_x + vv_y + wv_z = -\frac{1}{\rho}p_y - fu + \mathcal{Y}$$

$$u_t + uu_x + vu_y + wu_z = -\frac{1}{\rho}p_x + fv + \mathcal{X}$$

$$\left.\right\} \quad (I.20)$$

Differentiate the y component with respect to x and the x component with respect to y. Assume that ρ_x, ρ_y, w_x, and w_y are small enough to be neglected, and notice that f_x is zero but f_y is not. Subtract the x component equation from the y component and collect terms. The result is

$$\xi_t + u\xi_x + v\xi_y + w\xi_z + vf_y + (\xi + f)(u_x + v_y) = \mathcal{Y}_x - \mathcal{X}_y$$

Since

$$\frac{D}{Dt}(f) = vf_y$$

(see derivation of acceleration), we can write

$$\frac{D}{Dt}(\xi + f) + (\xi + f)(u_x + v_y) = \mathcal{Y}_x - \mathcal{X}_y \qquad (I.43)$$

The terms ξ and $\xi + f$ are defined as relative vorticity and absolute vorticity, respectively.

A useful special case of the vorticity equation is to consider a layer of constant density. First integrate the equation of continuity over a layer of thickness Z:

$$\int_{-Z}^{0} (u_x + v_y)\, dz = -\int_{-Z}^{0} w_z\, dz$$

$$(u_x + v_y)Z = -\int_{-Z}^{0} \frac{\partial}{\partial z}\left(\frac{dz}{dt}\right) dz = -\frac{dZ}{dt}$$

Then substitute into Eq. (I.43).

$$\frac{D}{Dt}(\xi+f) - \frac{1}{Z}(\xi+f)\frac{DZ}{Dt} = (\mathcal{Y}_x - \mathcal{X}_y)$$

It is assumed that all terms in parentheses are averaged over the layer Z. The thickness Z is only a function of time:

$$Z\frac{D}{Dt}\left(\frac{\xi+f}{Z}\right) = \mathcal{Y}_x - \mathcal{X}_y$$

or

$$\frac{D}{Dt}\left(\frac{\xi+f}{Z}\right) = 0 \tag{I.44}$$

in the absence of friction. Equation (I.44) is the form of the vorticity equation used as a starting point in the problem discussed in Chapter 7 on page 131. The term inside the parentheses is called *potential vorticity*.

Small-Amplitude Wave Equation

We start with the equation of motion, Eq. (I.21), with only the pressure-gradient force and gravity; that is, we ignore friction and the Coriolis term. In vector notation that is

$$\frac{\partial \mathbf{V}}{\partial t} + (\mathbf{V} \cdot \nabla) \mathbf{V} = -\frac{1}{\rho}\nabla p + \mathbf{g} \tag{I.45}$$

We assume that the fluid is homogeneous and incompressible and introduce the continuity equation, Eq. (I.3):

$$\nabla \cdot \mathbf{V} = 0$$

We also assume that the motion is irrotational:

$$\nabla \times \mathbf{V} = 0$$

Because of the assumption of irrotational motion, we can introduce a scalar velocity potential ϕ, which is defined as

$$\mathbf{V} = \nabla\phi, \qquad u = \phi_x, \qquad v = \phi_y, \qquad w = \phi_z$$

and because of the continuity equation, Eq. (I.3),

$$\nabla^2\phi = 0$$

For two-dimensional flow,

$$\nabla^2\phi = \phi_{xx} + \phi_{zz} = 0 \tag{I.46}$$

The x and z components of Eq. (I.45) can be written in terms of ϕ as

$$\phi_{xt} + \phi_x\phi_{xx} + \phi_z\phi_{xz} = -\frac{1}{\rho}\,p_x$$

$$\phi_{zt} + \phi_x\phi_{xz} + \phi_z\phi_{zz} = -\frac{1}{\rho}\,p_z - g$$

It can be shown that the above equations are the x and z derivatives of the following:

$$\phi_t + \tfrac{1}{2}\,(\phi_x{}^2 + \phi_z{}^2) = -\frac{p}{\rho} - gz + \text{constant}$$

or in terms of velocity

$$\phi_t + \tfrac{1}{2}\,\mathbf{V}^2 = -\frac{p}{\rho} - gz + \text{constant} \tag{I.47}$$

Note that this integral form of the equation of motion is a form of the Bernoulli equation, which is perhaps most familiar for the case of steady-state flow:

$$\tfrac{1}{2}\,\mathbf{V}^2 + \frac{p}{\rho} + gz = \text{constant} \tag{I.48}$$

We can now proceed directly to the small-amplitude wave equation (see Figure I.6 for some of the following definitions). We start with

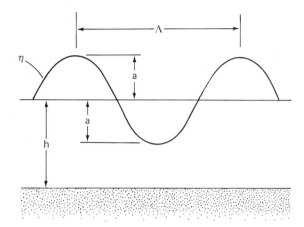

Figure I.6.

$$\phi_{xx} + \phi_{zz} = 0 \tag{I.46}$$

$$\frac{p}{\rho} + \phi_t + \tfrac{1}{2}V^2 + gz = c \tag{I.47}$$

and the following boundary condition on the bottom and the free surface. For the bottom at $z = -h$,

$$\phi_z = 0 \tag{I.49}$$

which simply means that the bottom is horizontal and there is no flow, normal to it. At the free surface $z = \eta$, we have the following condition:

$$\phi_z = \eta_t + \phi\check{\eta}_x$$
$$\phi_t + \tfrac{1}{2}\check{V}^2 + g\eta = 0$$

The first condition implies that the free surface is a streamline. The second indicates that there is no pressure gradient or other stress across the free surface and that the pressure of the overlying air can be neglected. If one further assumes that the wave amplitude is much less than the wavelength, the two terms with the check marks above them are sufficiently small to be neglected. Under these conditions, the two boundary conditions for the free surface can be combined into a single boundary-value equation:

$$\phi_{tt} + g\phi_z = 0 \tag{I.50}$$

Having our boundary conditions, we next specify a specific solution to Eq. (I.46). One possible solution is a sinusoidal wave with amplitude a that is a function of depth:

$$\phi(x, z, t) = a(z) \sin(\kappa x - \sigma t) \tag{I.51}$$

In terms of Eq. (I.46), this is

$$a_{zz} - \kappa^2 a = 0$$

a solution of which is

$$a(z) = A \cosh[\kappa(z + h)] + B \sinh[\kappa(z + h)]$$

and using Eq. (I.51) we get

$$\phi = (A \cosh[\kappa(z + h)] + B \sinh[\kappa(z + h)]) \sin(\kappa x - \sigma t) \tag{I.52}$$

We next apply the boundary condition of Eqs. (I.49) and (I.50) to the general solution of Eq. (I.46). The bottom boundary condition requires that

$$B = 0$$

The free surface boundary condition requires that

$$g\kappa \sinh \kappa h \sin(\kappa x - \sigma t) = \sigma^2 \cosh(\kappa h) \sin(\kappa x - \sigma t)$$

Since the phase velocity of the wave can be written as

$$C = \frac{\sigma}{\kappa}$$

then

$$C^2 = \frac{g}{\kappa} \tanh \kappa h \tag{I.53}$$

Group Velocity

Consider the superposition of two sets of sinusoidal waves of the same amplitude but slightly differing wavelength. The equation of the free surface will be the sum of these waves:

$$\eta = a \sin (\kappa x - \sigma t) + a \sin (\kappa' x - \sigma' t)$$

which can be written

$$\eta = 2a \cos \left[\tfrac{1}{2} (\kappa - \kappa')x - \tfrac{1}{2} (\sigma - \sigma')t\right] \sin \tfrac{1}{2} \left[(\kappa + \kappa')x - \tfrac{1}{2}(\sigma + \sigma')t\right]$$

If $\kappa - \kappa'$ is small, then the cosine term varies slowly with x, and the resulting free surface is a series of sine waves whose amplitude varies gradually from 0 to $2a$. The distance between two successive maxima or minima is $2\pi/(\kappa - \kappa')$, and the time between the passage of successive crests or troughs is $2\pi/(\sigma - \sigma')$. The phase velocity of this newly defined wave on the free surface is the *group velocity* of the combined waves. This velocity is

$$\mathcal{V} = \frac{\sigma - \sigma'}{\kappa - \kappa'}$$

which in the limit becomes

$$\mathcal{V} = \frac{d\sigma}{d\kappa}$$

$$= \frac{d}{d\kappa} (\kappa C)$$

(I.54)

For deep and shallow water waves,

$$\mathcal{V}_d = \tfrac{1}{2}\left(\frac{g}{\kappa}\right)^{1/2} = \tfrac{1}{2} C_d$$

(I.55)

$$\mathcal{V}_s = (gh)^{1/2} = C_s$$

(I.56)

Residence Time

Assume complete mixing of the incoming flow T as in Figure 4.3c. We assume that the incoming flow carries η units/cm³ of whatever it is we are measuring. η could be grams of salt/cm³, or dissolved oxygen, or zooplankton density, or anything that can be measured in numbers per unit volume. The rate at which new material is being added to the volume is ηT.

At any time t, there are N units of this new material in the volume V. Thus the rate at which the new material is being flushed out is $(N/V)\, T$. The rate of change is

$$\frac{dN}{dt} = \eta T - \frac{N}{V} T$$

Integrating over time,

$$\int_0^N \frac{dN}{\left(\eta - \dfrac{N}{V}\right)} = \int_0^t T\, dt$$

$$\ln\left(\eta - \frac{N}{V}\right) - \ln \eta = -\frac{T}{V} t$$

$$N = \eta V(1 - e^{-(T/V)t})$$

When $t = (V/T)$,

$$N = \eta V\left(1 - \frac{1}{e}\right)$$

Thus a residence time $\Xi = V/T$ is defined as the time it takes to replace all but $1/e$ (or 37%) of the original material.

appendix II

TABLES OF SOME PHYSICAL PROPERTIES OF SEAWATER

Computing Sigma-*t* If One Knows the Temperature and Salinity

Example of Computation:
  Given a temperature of 15.70°C and a salinity of 36.47⁰/00 compute the σ_t value.

  1. Select the salinity interval of 30.00 to 39.99⁰/00.
  2. In column one, find the temperature interval in which 15.70 falls (always use the *lower* limit of the interval). The *lower* limit is 15.69°C.
  3. Entering column one at 15.69°C read the corresponding value of 22.00 in column two. This is the correct σ_t value for the base of the salinity interval, that is, for a salinity of 30.00⁰/00, and temperature of 15.69°C.
  4. To find the correct σ_t value for the given salinity of 36.47⁰/00, multiply the designated f-factor in column three (.7680) by the last three digits of the given salinity (6.47), observing decimal places, and add the value obtained to the base value 22.00.
  5. Round the value obtained (26.96896) to two decimal places.   **Answer 26.97.**

Thus: Given 15.70°C and 36.47⁰/00 S.
  From table for Salinity 30.00 to 39.99⁰/00, enter column one at lower limit of temperature interval (15.69):

Obtain base		f-factor		last three	
value in		of column		digits of	
column two	+	three	×	given S	=
22.00		.7680		6.47	

  26.968960 (round to two decimal places)   **Answer 26.97**

Table II-1 Computing Sigma-*t* If One Knows the Temperature and Salinity*

DENSITY (σ_t)

Salinity 20.00°/oo to 29.99°/oo

T. °C.	σ_t	f	T. °C.	σ_t	f	T. °C.	σ_t	f
12.06	15.01		14.43	14.60	.7670	16.52	14.19	
.12	.00		.49	.59		.57	.18	
.19	14.99					.62	.17	
.25	.98	.7710	14.54	14.58		.67	.16	
.31	.97		.59	.57		.72	.15	.7620
.37	.96		.65	.56		.76	.14	
.42	.95		.70	.55		.81	.13	
.49	.94		.75	.54	.7660	.86	.12	
			.81	.53		.91	.11	
12.55	14.93		.86	.52		.95	.10	
.61	.92		.91	.51				
.67	.91		.96	.50		17.00	14.09	
.73	.90					.05	.08	
.79	.89	.7700	15.01	14.49		.10	.07	
.85	.88		.07	.48		.14	.06	
.91	.87		.12	.47		.19	.05	
.97	.86		.17	.46		.24	.04	.7610
			.22	.45		.28	.03	
13.03	14.85		.27	.44	.7650	.33	.02	
.09	.84		.33	.43		.38	.01	
.14	.83		.38	.42		.42	.00	
.20	.82		.43	.41		.47	13.99	
.26	.81	.7690	.48	.40				
.32	.80					17.52	13.98	
.38	.79		15.53	14.39		.56	.97	
.43	.78		.58	.38		.61	.96	
.49	.77		.63	.37		.65	.95	
			.68	.36		.70	.94	
13.55	14.76		.73	.35		.75	.93	.7600
.60	.75		.78	.34	.7640	.79	.92	
.66	.74		.83	.33		.84	.91	
.72	.73		.88	.32		.88	.90	
.77	.72	.7680	.93	.31		.93	.89	
.83	.71		.98	.30		.97	.88	
.88	.70							
.94	.69		16.03	14.29		18.02	13.87	
.99	.68		.08	.28		.06	.86	
			.13	.27		.11	.85	
14.05	14.67		.18	.26		.15	.84	
.10	.66		.23	.25		.20	.83	
.16	.65		.28	.24	.7630	.24	.82	.7590
.21	.64	.7670	.33	.23		.29	.81	
.27	.63		.38	.22		.33	.80	
.32	.62		.43	.21		.38	.79	
.38	.61		.47	.20		.42	.78	

a

298

Table II-1 (continued)

DENSITY (σ_t)

Salinity 20.00°/oo to 29.99°/oo

T. °C.	σ_t	f	T. °C.	σ_t	f	T. °C.	σ_t	f
18.47	13.77	.7590	20.22	13.36		21.90	12.94	
			.26	.35		.94	.93	.7540
18.51	13.76		.30	.34		.98	.92	
.55	.75		.34	.33	.7560			
.60	.74		.38	.32		22.03	12.91	
.64	.73		.42	.31		.06	.90	
.69	.72		.47	.30		.09	.89	
.73	.71					.13	.88	.7540
.77	.70	.7580	20.51	13.29		.17	.87	
.82	.69		.55	.28		.21	.86	
.86	.68		.59	.27		.25	.85	
.91	.67		.63	.26				
.95	.66		.67	.25		22.29	12.84	
.99	.65		.71	.24		.32	.83	
			.75	.23	.7560	.36	.82	
19.04	13.64		.79	.22		.40	.81	
.08	.63		.83	.21		.44	.80	
.12	.62		.87	.20		.48	.79	
.17	.61		.91	.19		.51	.78	
.21	.60		.95	.18		.55	.77	
.25	.59	.7570	.99	.17		.59	.76	
.29	.58					.63	.75	.7530
.34	.57		21.03	13.16		.67	.74	
.38	.56		.07	.15		.70	.73	
.42	.55		.11	.14		.74	.72	
.46	.54		.15	.13		.78	.71	
			.19	.12		.82	.70	
19.51	13.53		.23	.11		.85	.69	
.55	.52		.27	.10	.7550	.89	.68	
.59	.51		.31	.09		.93	.67	
.63	.50		.35	.08		.97	.66	
.68	.49		.39	.07				
.72	.48		.43	.06		23.00	12.65	
.76	.47	.7560	.47	.05		.04	.64	
.80	.46					.08	.63	
.84	.45		21.51	13.04		.12	.62	
.89	.44		.55	.03		.15	.61	
.93	.43		.59	.02		.19	.60	
.97	.42		.63	.01		.23	.59	.7520
			.67	.00		.27	.58	
20.01	13.41		.71	12.99	.7540	.30	.57	
.05	.40		.74	.98		.34	.56	
.10	.39	.7560	.78	.97		.38	.55	
.14	.38		.82	.96		.41	.54	
.18	.37		.86	.95		.45	.53	

b

Table II-1 (continued)

DENSITY (σ_t)

Salinity 20.00°/oo to 29.99°/oo

T. °C.	σ_t	f	T. °C.	σ_t	f	T. °C.	σ_t	f
23.49	12.52	.7520	24.96	12.11	.7500	26.37	11.70	
			.99	.10		.40	.69	
23.52	12.51					.44	.68	.7490
.56	.50		25.03	12.09		.47	.67	
.60	.49		.06	.08				
.63	.48		.10	.07		26.50	11.66	
.67	.47		.13	.06		.54	.65	
.71	.46		.17	.05		.57	.64	
.74	.45	.7510	.20	.04		.60	.63	
.78	.44		.23	.03		.63	.62	
.82	.43		.27	.02	.7500	.66	.61	
.85	.42		.31	.01		.70	.60	
.89	.41		.34	.00		.74	.59	.7480
.92	.40		.38	11.99		.77	.58	
.96	.39		.41	.98		.80	.57	
			.45	.97		.83	.56	
24.00	12.38		.48	.96		.87	.55	
.03	.37					.90	.54	
.07	.36		25.52	11.95		.94	.53	
.11	.35		.55	.94		.97	.52	
.14	.34		.59	.93				
.18	.33		.62	.92		27.00	11.51	
.21	.32		.65	.91		.04	.50	
.25	.31	.7510	.69	.90		.07	.49	
.29	.30		.72	.89		.10	.48	
.32	.29		.76	.88	.7490	.13	.47	
.36	.28		.79	.87		.17	.46	
.39	.27		.83	.86		.20	.45	
.43	.26		.86	.85		.23	.44	.7480
.46	.25		.89	.84		.27	.43	
			.93	.83		.30	.42	
24.50	12.24		.96	.82		.33	.41	
.54	.23					.36	.40	
.57	.22		26.00	11.81		.40	.39	
.61	.21		.03	.80		.43	.38	
.64	.20		.06	.79		.46	.37	
.68	.19		.10	.78				
.71	.18		.13	.77		27.50	11.36	
.75	.17	.7500	.17	.76	.7490	.53	.35	
.78	.16		.20	.75		.56	.34	
.82	.15		.23	.74		.59	.33	.7470
.85	.14		.27	.73		.63	.32	
.88	.13		.30	.72		.66	.31	
.92	.12		.34	.71		.69	.30	

c

Table II-1 (continued)
DENSITY (σ_t)

Salinity 20.00°/oo to 29.99°/oo

T. °C.	σ_t	f
27.72	11.29	
.75	.28	
.78	.27	
.82	.26	
.85	.25	.7470
.88	.24	
.92	.23	
.95	.22	
.98	.21	
28.01	11.20	
.05	.19	
.08	.18	
.11	.17	
.14	.16	
.17	.15	
.21	.14	
.24	.13	
.27	.12	.7460
.30	.11	
.33	.10	
.36	.09	
.40	.08	
.43	.07	
.46	.06	
.49	.05	
28.52	11.04	
.56	.03	
.59	.02	
.62	.01	
.65	.00	
.68	10.99	
.71	.98	.7460
.75	.97	
.78	.96	
.81	.95	
.84	.94	
.87	.93	

T. °C.	σ_t	f
28.90	10.92	
.93	.91	.7460
.97	.90	
29.00	10.89	
.03	.88	
.06	.87	
.09	.86	
.12	.85	
.15	.84	
.18	.83	
.21	.82	
.25	.81	.7450
.28	.80	
.31	.79	
.34	.78	
.37	.77	
.40	.76	
.43	.75	
.46	.74	
.49	.73	
29.52	10.72	
.56	.71	
.59	.70	
.62	.69	
.65	.68	
.68	.67	
.71	.66	
.74	.65	
.77	.64	.7450
.80	.63	
.83	.62	
.86	.61	
.89	.60	
.92	.59	
.96	.58	
.99	.57	

d

Table II-1 (continued)

DENSITY (σ_t)

Salinity 30.00°/oo to 39.99°/oo

T. °C.	σ_t	f
-2.00	24.15	.8120
-1.75	.14	
-1.13	24.13	.8100
-0.71	24.12	.8090
-0.37	24.11	.8070
-0.08	.10	
0.18	24.09	.8050
0.42	.08	
0.64	24.07	.8040
0.85	.06	
1.05	24.05	.8020
.24	.04	
.41	.03	
1.58	24.02	.8010
.75	.01	
.91	24.00	
2.06	23.99	.8000
.21	.98	
.35	.97	
2.50	23.96	
.63	.95	.7980
.77	.94	
.90	.93	
3.03	23.92	
.15	.91	.7970
.27	.90	
.40	.89	
3.51	23.88	
.62	.87	
.74	.86	.7950
.86	.85	
.97	.84	

T. °C.	σ_t	f
4.07	23.83	
.18	.82	.7940
.29	.81	
.40	.80	
4.50	23.79	
.60	.78	
.70	.77	.7930
.80	.76	
.90	,75	
5.00	23.74	
.09	.73	
.19	.72	.7910
.27	.71	
.37	.70	
.46	.69	
5.56	23.68	
.65	.67	
.73	.66	.7900
.82	.65	
.91	.64	
6.00	23.63	
.08	.62	
.17	.61	.7890
.25	.60	
.34	.59	
.42	.58	
6.50	23.57	
.59	.56	
.67	.55	
.75	.54	.7880
.83	.53	
.91	.52	
.99	.51	
7.06	23.50	
.14	.49	.7860
.22	.48	
.30	.47	

T. °C.	σ_t	f
7.37	23.46	.7860
.45	.45	
7.52	23.44	
.60	.43	
.67	.42	
.75	.41	.7850
.82	.40	
.89	.39	
.96	.38	
8.04	23.37	
.11	.36	
.18	.35	
.25	.34	.7840
.32	.33	
.39	.32	
.46	.31	
8.53	23.30	
.60	.29	
.67	.28	
.74	.27	.7830
.80	.26	
.87	.25	
.94	.24	
9.01	23.23	
.07	.22	
.14	.21	
.20	.20	
.27	.19	.7820
.34	.18	
.40	.17	
.47	.16	
9.53	23.15	
.59	.14	
.66	.13	
.72	.12	.7810
.79	.11	
.85	.10	
.91	.09	

e

Table II-1 (continued)

DENSITY (σ_t)

Salinity 30.00°/oo to 39.99°/oo

T. °C.	σ_t	f
9.97	23.08	.7810
10.04	23.07	
.10	.06	
.16	.05	
.22	.04	
.28	.03	.7790
.34	.02	
.40	.01	
.47	.00	
10.53	22.99	
.59	.98	
.65	.97	
.71	.96	
.77	.95	.7780
.82	.94	
.88	.93	
.94	.92	
11.00	22.91	
.06	.90	
.12	.89	
.18	.88	
.23	.87	.7770
.29	.86	
.35	.85	
.41	.84	
.46	.83	
11.52	22.82	
.58	.81	
.63	.80	
.69	.79	
.74	.78	.7760
.80	.77	
.86	.76	
.91	.75	
.97	.74	
12.02	22.73	
.08	.72	
.13	.71	.7750
.19	.70	

T. °C.	σ_t	f
12.24	22.69	
.30	.68	
.35	.67	.7750
.40	.66	
.46	.65	
12.51	22.64	
.57	.63	
.62	.62	
.67	.61	
.73	.60	
.78	.59	.7740
.83	.58	
.88	.57	
.94	.56	
.99	.55	
13.04	22.54	
.09	.53	
.15	.52	
.20	.51	
.25	.50	.7730
.30	.49	
.35	.48	
.40	.47	
.46	.46	
13.51	22.45	
.56	.44	
.61	.43	
.66	.42	
.71	.41	
.76	.40	.7720
.81	.39	
.86	.38	
.91	.37	
.96	.36	
14.01	22.35	
.06	.34	
.11	.33	
.16	.32	.7710
.21	.31	
.26	.30	

T. °C.	σ_t	f
14.31	22.29	
.36	.28	
.40	.27	.7710
.45	.26	
14.50	22.25	
.55	.24	
.60	.23	
.65	.22	
.70	.21	
.74	.20	.7700
.79	.19	
.84	.18	
.89	.17	
.94	.16	
.98	.15	
15.03	22.14	
.08	.13	
.13	.12	
.17	.11	
.22	.10	
.27	.09	.7690
.31	.08	
.36	.07	
.41	.06	
.45	.05	
15.50	22.04	
.55	.03	
.59	.02	
.64	.01	
.69	.00	
.73	21.99	.7680
.78	.98	
.82	.97	
.87	.96	
.92	.95	
.96	.94	
16.01	21.93	
.05	.92	
.10	.91	.7680
.14	.90	

f

Table II-1 (continued)

DENSITY (σ_t)

Salinity 30.00°/oo to 39.99°/oo

T.°C.	σ_t	f	T.°C.	σ_t	f	T.°C.	σ_t	f
16.19	21.89		18.01	21.47		19.72	21.05	
.23	.88		.06	.46		.76	.04	
.28	.87		.10	.45		.80	.03	
.32	.86	.7680	.14	.44		.84	.02	.7620
.37	.85		.18	.43		.88	.01	
.41	.84		.22	.42		.92	.00	
.46	.83		.27	.41	.7640	.96	20.99	
			.31	.40				
16.50	21.82		.35	.39		20.00	20.98	
.55	.81		.39	.38		.04	.97	
.59	.80		.43	.37		.08	.96	
.63	.79		.47	.36		.12	.95	
.68	.78					.15	.94	
.72	.77		18.51	21.35		.19	.93	
.77	.76	.7670	.55	.34		.23	.92	.7610
.81	.75		.60	.33		.27	.91	
.86	.74		.64	.32		.31	.90	
.90	.73		.68	.31		.35	.89	
.94	.72		.72	.30		.39	.88	
.99	.71		.76	.29	.7630	.43	.87	
			.80	.28		.46	.86	
17.03	21.70		.84	.27				
.07	.69		.88	.26		20.50	20.85	
.12	.68		.92	.25		.54	.84	
.16	.67		.96	.24		.58	.83	
.20	.66					.62	.82	
.25	.65	.7660	19.00	21.23		.66	.81	
.29	.64		.04	.22		.69	.80	
.33	.63		.08	.21		.73	.79	.7600
.38	.62		.13	.20		.77	.78	
.42	.61		.17	.19		.81	.77	
.46	.60		.21	.18		.85	.76	
			.25	.17	.7630	.89	.75	
17.51	21.59		.29	.16		.92	.74	
.55	.58		.33	.15		.96	.73	
.59	.57		.37	.14				
.63	.56		.41	.13		21.00	20.72	
.68	.55		.45	.12		.04	.71	
.72	.54		.49	.11		.07	.70	
.76	.53	.7650				.11	.69	
.80	.52		19.53	21.10		.15	.68	.7600
.85	.51		.57	.09		.19	.67	
.89	.50		.60	.08	.7620	.22	.66	
.93	.49		.64	.07		.26	.65	
.97	.48		.68	.06		.30	.64	

g

Table II-1 (continued)

DENSITY (σ_t)

Salinity 30.00°/oo to 39.99°/oo

T. °C.	σ_t	f	T. °C.	σ_t	f	T. °C.	σ_t	f
21.34	20.63		22.84	20.22		24.27	19.81	
.37	.62		.87	.21		.30	.80	
.41	.61	.7600	.91	.20	.7580	.34	.79	
.45	.60		.94	.19		.37	.78	.7560
.49	.59		.98	.18		.41	.77	
						.44	.76	
21.52	20.58		23.01	20.17		.48	.75	
.56	.57		.05	.16				
.60	.56		.09	.15		24.51	19.74	
.64	.55		.12	.14		.54	.73	
.67	.54		.16	.13		.58	.72	
.71	.53		.19	.12		.61	.71	
.75	.52	.7590	.23	.11		.65	.70	
.78	.51		.26	.10	.7570	.68	.69	
.82	.50		.30	.09		.71	.68	
.86	.49		.33	.08		.75	.67	.7550
.89	.48		.37	.07		.78	.66	
.93	.47		.40	.06		.81	.65	
.97	.46		.44	.05		.85	.64	
			.47	.04		.88	.63	
22.00	20.45					.92	.62	
.04	.44		23.51	20.03		.95	.61	
.08	.43		.54	.02		.98	.60	
.11	.42		.58	.01				
.15	.41		.61	.00		25.02	19.59	
.19	.40		.65	19.99		.05	.58	
.22	.39		.68	.98		.08	.57	
.26	.38	.7580	.72	.97		.12	.56	
.30	.37		.75	.96	.7560	.15	.55	
.33	.36		.79	.95		.18	.54	
.37	.35		.82	.94		.22	.53	
.40	.34		.86	.93		.25	.52	.7550
.44	.33		.89	.92		.28	.51	
.48	.32		.93	.91		.32	.50	
			.96	.90		.35	.49	
22.51	20.31					.38	.48	
.55	.30		24.00	19.89		.42	.47	
.59	.29		.03	.88		.45	.46	
.62	.28		.07	.87		.48	.45	
.66	.27	.7580	.10	.86				
.69	.26		.13	.85	.7560	25.52	19.44	
.73	.25		.17	.84		.55	.43	
.76	.24		.20	.83		.58	.42	.7540
.80	.23		.24	.82		.62	.41	

h

Table II-1 (continued)
DENSITY (σ_t)

Salinity 30.00°/oo to 39.99°/oo

T. °C.	σ_t	f
25.65	19.40	
.68	.39	
.71	.38	
.75	.37	
.78	.36	
.81	.35	.7540
.85	.34	
.88	.33	
.91	.32	
.94	.31	
.98	.30	
26.01	19.29	
.04	.28	
.08	.27	
.11	.26	
.14	.25	
.17	.24	
.21	.23	
.24	.22	.7530
.27	.21	
.30	.20	
.34	.19	
.37	.18	
.40	.17	
.43	.16	
.46	.15	
26.50	19.14	
.53	.13	
.56	.12	
.59	.11	
.63	.10	
.66	.09	
.69	.08	
.72	.07	
.75	.06	.7530
.79	.05	
.82	.04	
.85	.03	
.88	.02	
.91	.01	
.95	.00	
.98	18.99	

T. °C.	σ_t	f
27.01	18.98	
.04	.97	
.07	.96	
.11	.95	
.14	.94	
.17	.93	
.20	.92	
.23	.91	
.26	.90	.7520
.30	.89	
.33	.88	
.36	.87	
.39	.86	
.42	.85	
.45	.84	
.48	.83	
27.52	18.82	
.55	.81	
.58	.80	
.61	.79	
.64	.78	
.67	.77	
.70	.76	
.74	.75	
.77	.74	.7520
.80	.73	
.83	.72	
.86	.71	
.89	.70	
.92	.69	
.95	.68	
.98	.67	
28.02	18.66	
.05	.65	
.08	.64	
.11	.63	
.14	.62	
.17	.61	.7510
.20	.60	
.23	.59	
.26	.58	
.29	.57	

T. °C.	σ_t	f
28.32	18.56	
.36	.55	
.39	.54	
.42	.53	.7510
.45	.52	
.48	.51	
28.51	18.50	
.54	.49	
.57	.48	
.60	.47	
.63	.46	
.66	.45	
.69	.44	
.72	.43	.7510
.75	.42	
.78	.41	
.81	.40	
.85	.39	
.88	.38	
.91	.37	
.94	.36	
.97	.35	
29.00	18.34	
.03	.33	
.06	.32	
.09	.31	
.12	.30	
.15	.29	
.18	.28	
.21	.27	
.24	.26	.7510
.27	.25	
.30	.24	
.33	.23	
.36	.22	
.39	.21	
.42	.20	
.45	.19	
.48	.18	
29.51	18.17	.7500
.54	.16	

i

Table II-1 (continued)
DENSITY (σ_t)

Salinity $30.00^\circ/oo$ to $39.99^\circ/oo$

T. °C.	σ_t	f
29.57	18.15	
.60	.14	
.63	.13	
.66	.12	
.69	.11	
.72	.10	
.75	.09	
.78	.08	.7500
.81	.07	
.84	.06	
.87	.05	
.90	.04	
.93	.03	
.96	.02	
.99	.01	

i

*From U.S. Navy Hydrographic Office, 1956: Tables for the rapid computation of density and electrical conducting of seawater. Special Publication 11, Washington, D.C.

Table II-2 Conversion of Sigma-*t* to Thermosteric Anomaly*

Example:
Given, $\sigma_t = 26.32$.
From table $10^5 \Delta_{s,t} = 171.2$.

σ_t	.00	.01	.02	.03	.04	.05	.06	.07	.08	.09
16.0	1160.9	1159.9	1158.9	1158.0	1157.0	1156.0	1155.1	1154.1	1153.1	1152.2
16.1	1151.2	1150.2	1149.3	1148.3	1147.3	1146.4	1145.4	1144.4	1143.5	1142.5
16.2	1141.5	1140.5	1139.6	1138.6	1137.6	1136.7	1135.7	1134.7	1133.8	1132.8
16.3	1131.8	1130.9	1129.9	1128.9	1128.0	1127.0	1126.0	1125.1	1124.1	1123.1
16.4	1122.1	1121.2	1120.2	1119.2	1118.3	1117.3	1116.3	1115.4	1114.4	1113.4
16.5	1112.5	1111.5	1110.5	1109.6	1108.6	1107.6	1106.7	1105.7	1104.7	1103.8
16.6	1102.8	1101.8	1100.9	1099.9	1098.9	1098.0	1097.0	1096.0	1095.1	1094.1
16.7	1093.1	1092.1	1091.2	1090.2	1089.2	1088.3	1087.3	1086.3	1085.4	1084.4
16.8	1083.4	1082.5	1081.5	1080.5	1079.6	1078.6	1077.6	1076.7	1075.7	1074.7
16.9	1073.8	1072.8	1071.8	1070.9	1069.9	1068.9	1068.0	1067.0	1066.0	1065.1
17.0	1064.1	1063.1	1062.2	1061.2	1060.2	1059.3	1058.3	1057.3	1056.4	1055.4
17.1	1054.4	1053.5	1052.5	1051.5	1050.6	1049.6	1048.6	1047.7	1046.7	1045.7
17.2	1044.8	1043.8	1042.8	1041.9	1040.9	1039.9	1039.0	1038.0	1037.0	1036.1
17.3	1035.1	1034.1	1033.2	1032.2	1031.2	1030.3	1029.3	1028.3	1027.4	1026.4
17.4	1025.4	1024.5	1023.5	1022.5	1021.6	1020.6	1019.6	1018.7	1017.7	1016.7
17.5	1015.8	1014.8	1013.9	1012.9	1011.9	1011.0	1010.0	1009.0	1008.1	1007.1
17.6	1006.1	1005.2	1004.2	1003.2	1002.3	1001.3	1000.3	999.4	998.4	997.4
17.7	996.5	995.5	994.5	993.6	992.6	991.6	990.7	989.7	988.7	987.8
17.8	986.8	985.8	984.9	983.9	983.0	982.0	981.0	980.1	979.1	978.1
17.9	977.2	976.2	975.2	974.3	973.3	972.3	971.4	970.4	969.4	968.5
18.0	967.5	966.6	965.6	964.6	963.7	962.7	961.7	960.8	959.8	958.8
18.1	957.9	956.9	955.9	955.0	954.0	953.1	952.1	951.1	950.2	949.2
18.2	948.2	947.3	946.3	945.3	944.4	943.4	942.4	941.5	940.5	939.5
18.3	938.6	937.6	936.7	935.7	934.7	933.8	932.8	931.8	930.9	929.9
18.4	928.9	928.0	927.0	926.0	925.1	924.1	923.2	922.2	921.2	920.3
18.5	919.3	918.3	917.4	916.4	915.4	914.5	913.5	912.6	911.6	910.6
18.6	909.7	908.7	907.7	906.8	905.8	904.8	903.9	902.9	902.0	901.0
18.7	900.0	899.1	898.1	897.1	896.2	895.2	894.2	893.3	892.3	891.4
18.8	890.4	889.4	888.5	887.5	886.5	885.6	884.6	883.6	882.7	881.7
18.9	880.8	879.8	878.8	877.9	876.9	875.9	875.0	874.0	873.0	872.1
19.0	871.1	870.2	869.2	868.2	867.3	866.3	865.3	864.4	863.4	862.5
19.1	861.5	860.5	859.6	858.6	857.6	856.7	855.7	854.8	853.8	852.8
19.2	851.9	850.9	849.9	849.0	848.0	847.0	846.1	845.1	844.2	843.2
19.3	842.2	841.3	840.3	839.4	838.4	837.4	836.5	835.5	834.5	833.6
19.4	832.6	831.7	830.7	829.7	828.8	827.8	826.8	825.9	824.9	824.0
19.5	823.0	822.0	821.1	820.1	819.1	818.2	817.2	816.3	815.3	814.3
19.6	813.4	812.4	811.5	810.5	809.5	808.6	807.6	806.6	805.7	804.7
19.7	803.8	802.8	801.8	800.9	799.9	798.9	798.0	797.0	796.1	795.1
19.8	794.1	793.2	792.2	791.3	790.3	789.3	788.4	787.4	786.4	785.5
19.9	784.5	783.6	782.6	781.6	780.7	779.7	778.8	777.8	776.8	775.9
20.0	774.9	773.9	773.0	772.0	771.1	770.1	769.1	768.2	767.2	766.3
20.1	765.3	764.3	763.4	762.4	761.5	760.5	759.5	758.6	757.6	756.7
20.2	755.7	754.7	753.8	752.8	751.8	750.9	749.9	749.0	748.0	747.0
20.3	746.1	745.1	744.2	743.2	742.2	741.3	740.3	739.4	738.4	737.4
20.4	736.5	735.5	734.6	733.6	732.6	731.7	730.7	729.8	728.8	727.8

a

Table II-2 (continued)

σ_t	.00	.01	.02	.03	.04	.05	.06	.07	.08	.09
20.5	726. 9	725. 9	725. 0	724. 0	723. 0	722. 1	721. 1	720. 2	719. 2	718. 2
20.6	717. 3	716. 3	715. 4	714. 4	713. 4	712. 5	711. 5	710. 6	709. 6	708. 6
20.7	707. 7	706. 7	705. 8	704. 8	703. 8	702. 9	701. 9	701. 0	700. 0	699. 0
20.8	698. 1	697. 1	696. 2	695. 2	694. 2	693. 3	692. 3	691. 4	690. 4	689. 4
20.9	688. 5	687. 5	686. 6	685. 6	684. 6	683. 7	682. 7	681. 8	680. 8	679. 8
21.0	678. 9	677. 9	677. 0	676. 0	675. 1	674. 1	673. 1	672. 2	671. 2	670. 3
21.1	669. 3	668. 3	667. 4	666. 4	665. 4	664. 5	663. 5	662. 6	661. 6	660. 7
21.2	659. 7	658. 7	657. 8	656. 8	655. 9	654. 9	654. 0	653. 0	652. 0	651. 1
21.3	650. 1	649. 2	648. 2	647. 2	646. 3	645. 3	644. 4	643. 4	642.. 5	641. 5
21.4	640. 5	639. 6	638. 6	637. 7	636. 7	635. 7	634. 8	633. 8	632. 9	631. 9
21.5	630. 9	630. 0	629. 0	628. 1	627. 1	626. 2	625. 2	624. 2	623. 3	622. 3
21.6	621. 4	620. 4	619. 5	618. 5	617. 5	616. 6	615. 6	614. 7	613. 7	612. 7
21.7	611. 8	610. 8	609. 9	608. 9	608. 0	607. 0	606. 0	605. 1	604. 1	603. 2
21.8	602. 2	601. 2	600. 3	599. 3	598. 4	597. 4	596. 5	595. 5	595. 5	593. 6
21.9	592. 6	591. 7	590. 7	589. 8	588. 8	587. 8	586. 9	585. 9	585. 0	584. 0
22.0	583. 1	582. 1	581. 1	580. 2	579. 2	578. 3	577. 3	576. 4	575. 4	574. 4
22.1	573. 5	572. 5	571. 6	570. 6	569. 7	568. 7	567. 7	566. 8	565. 8	564. 9
22.2	563. 9	563. 0	562. 0	561. 0	560. 1	559. 1	558. 2	557. 2	556. 3	555. 3
22.3	554. 3	553. 4	552. 4	551. 5	550. 5	549. 6	548. 6	547. 6	546. 7	545. 7
22.4	544. 8	543. 8	542. 9	541. 9	540. 9	540. 0	539. 0	538. 1	537. 1	536. 2
22.5	535. 2	534. 3	533. 3	532. 3	531. 4	530. 4	529. 5	528. 5	527. 6	526. 6
22.6	525. 6	524. 7	523. 7	522. 8	521. 8	520. 9	519. 9	519. 0	518. 0	517. 0
22.7	516. 1	515. 1	514. 2	513. 3	512. 3	511. 3	510. 3	509. 4	508. 4	507. 5
22.8	506. 5	505. 6	504. 6	503. 7	502. 7	501. 7	500. 8	499. 8	498. 9	497. 9
22.9	497. 0	496. 0	495. 1	494. 1	493. 1	492. 2	491. 2	490. 3	489. 3	488. 4
23.0	487. 4	486. 5	485. 5	484. 5	483. 6	482. 6	481. 7	480. 7	479. 8	478. 8
23.1	477. 9	476. 9	475. 9	475. 0	474. 0	473. 1	472. 1	471. 2	470. 2	469. 3
23.2	468. 3	467. 3	466. 4	465. 4	464. 5	463. 5	462. 6	461. 6	460. 7	459. 7
23.3	458. 7	457. 8	456. 8	455. 9	454. 9	454. 0	453. 0	452. 1	451. 1	450. 2
23.4	449. 2	448. 2	447. 3	446. 3	445. 4	444. 4	443. 5	442. 5	441. 6	440. 6
23.5	439. 7	438. 7	437. 7	436. 8	435. 8	434. 9	433. 9	433. 0	432. 0	431. 1
23.6	430. 1	429. 2	428. 2	427. 2	426. 3	425. 3	424. 4	423. 4	422. 5	421. 5
23.7	420. 6	419. 6	418. 7	417. 7	416. 7	415. 8	414. 8	413. 9	412. 9	412. 0
23.8	411. 0	410. 1	409. 1	408. 2	407. 2	406. 3	405. 3	404. 3	403. 4	402. 4
23.9	401. 5	400. 5	399. 6	398. 6	397. 7	396. 7	395. 8	394. 8	393. 9	392. 9
24.0	391. 9	391. 0	390. 0	389. 1	388. 1	387. 2	386. 2	385. 3	384. 3	383. 4
24.1	382. 4	381. 5	380. '5	379. 6	378. 6	377. 6	376. 7	375. 7	374. 8	373. 8
24.2	372. 9	371. 9	371. 0	370. 0	369. 1	368. 1	367. 2	366. 2	365. 3	364. 3
24.3	363. 3	362. 4	361. 4	360. 5	359. 5	358. 6	357. 6	356. 7	355. 7	354. 8
24.4	353. 8	352. 9	351. 9	351. 0	350. 0	349. 0	348. 1	347. 1	346. 2	345. 2
24.5	344. 3	343. 3	342. 4	341. 4	340. 5	339. 5	338. 6	337. 6	336. 7	335. 7
24.6	334. 8	333. 8	332. 9	331. 9	330. 9	330. 0	329. 0	328. 1	327. 1	326. 2
24.7	325. 2	324. 3	323. 3	322. 4	321. 4	320. 5	319. 5	318. 6	317. 6	316. 7
24.8	315. 7	314. 8	313. 8	312. 9	311. 8	311. 0	310. 0	309. 0	308. 1	307. 1
24.9	306. 2	305. 2	304. 3	303. 3	302. 4	301. 4	300. 5	299. 5	298. 6	297. 6
25.0	296. 7	295. 7	294. 8	293. 8	292. 9	291. 9	291. 0	290. 0	289. 1	288. 1
25.1	287. 2	286. 2	285. 3	284. 3	283. 3	282. 4	281. 4	280. 5	279. 5	278. 6
25.2	277. 6	276. 7	275. 7	274. 8	273. 8	272. 9	271. 9	271. 0	270. 0	269. 1
25.3	268. 1	267. 2	266. 2	265. 3	264. 3	263. 4	262. 4	261. 5	260. 5	259. 6
25.4	258. 6	257. 7	256. 7	255. 8	254. 8	253. 9	252. 9	252. 0	251. 0	250. 1
25.5	249. 1	248. 2	247. 2	246. 3	245. 3	244. 3	243. 4	242. 4	241. 5	240. 5
25.6	239. 6	238. 6	237. 7	236. 7	235. 8	234. 8	233. 9	232. 9	232. 0	231. 0
25.7	230. 1	229. 1	228. 2	227. 2	226. 3	225. 3	224. 4	223. 4	222. 5	221. 5
25.8	220. 6	219. 6	218. 7	217. 7	216. 8	215. 8	214. 9	213. 9	213. 0	212. 0
25.9	211. 1	210. 1	209. 2	208. 2	207. 3	206. 3	205. 4	204. 4	203. 5	202. 5

b

Table II-2 (continued)

σ_t	.00	.01	.02	.03	.04	.05	.06	.07	.08	.09
26.0	201. 6	200. 6	199. 7	198. 7	197. 8	196. 8	195. 9	194. 9	194. 0	193. 0
26.1	192. 1	191. 1	190. 2	189. 2	188. 3	187. 3	186. 4	185. 4	184. 5	183. 5
26.2	182. 6	181. 6	180. 7	179. 7	178. 8	177. 8	176. 9	175. 9	175. 0	174. 0
26.3	173. 1	172. 1	171. 2	170. 2	169. 3	168. 3	167. 4	166. 4	165. 5	164. 5
26.4	163. 6	162. 7	161. 7	160. 8	159. 8	158. 9	157. 9	157. 0	156. 0	155. 1
26.5	154. 1	153. 2	152. 2	151. 3	150. 3	149. 4	148. 4	147. 5	146. 5	145. 6
26.6	144. 6	143. 7	142. 7	141. 8	140. 8	139. 9	138. 9	138. 0	137. 0	136. 1
26.7	135. 1	134. 2	133. 2	132. 3	131. 3	130. 4	129. 4	128. 5	127. 5	126. 6
26.8	125. 6	124. 7	123. 7	122. 8	121. 9	120. 9	120. 0	119. 0	118. 1	117. 1
26.9	116. 2	115. 2	114. 3	113. 3	112. 4	111. 4	110. 5	109. 5	108. 6	107. 6
27.0	106. 7	105. 7	104. 8	103. 8	102. 9	101. 9	101. 0	100. 0	99. 1	98. 1
27.1	97. 2	96. 3	95. 3	94. 4	93. 4	92. 5	91. 5	90. 6	89. 6	88. 7
27.2	87. 7	86. 8	85. 8	84. 9	83. 9	83. 0	82. 0	81. 1	80. 1	79. 2
27.3	78. 2	77. 3	76. 3	75. 4	74. 5	73. 5	72. 6	71. 6	70. 7	69. 7
27.4	68. 8	67. 8	66. 9	65. 9	65. 0	64. 0	63. 1	62. 1	61. 2	60. 2
27.5	59. 3	58. 3	57. 4	56. 5	55. 5	54. 6	53. 6	52. 7	51. 7	50. 8
27.6	49. 8	48. 9	47. 9	47. 0	46. 0	45. 1	44. 1	43. 2	42. 3	41. 3
27.7	40. 4	39. 4	38. 5	37. 5	36. 6	35. 6	34. 7	33. 7	32. 8	31. 8
27.8	30. 9	29. 9	29. 0	28. 1	27. 1	26. 2	25. 2	24. 3	23. 3	22. 4
27.9	21. 4	20. 5	19. 5	18. 6	17. 6	16. 7	15. 7	14. 8	13. 9	12. 9
28.0	12. 0	11. 0	10. 1	9. 1	8. 2	7. 2	6. 3	5. 3	4. 4	3. 4
28.1	2. 5	1. 6	0. 6	−0. 3	−1. 3	−2. 2	−3. 2	−4. 1	−5. 1	−6. 0
28.2	−7. 0	−7. 9	−8. 9	−9. 8	−10. 8	−11. 7	−12. 6	−13. 6	−14. 5	−15. 5
28.3	−16. 4	−17. 4	−18. 3	−19. 3	−20. 2	−21. 2	−22. 1	−23. 0	−24. 0	−24. 9
28.4	−25. 9	−26. 8	−27. 8	−28. 7	−29. 7	−30. 6	−31. 6	−32. 5	−33. 4	−34. 4
28.5	−35. 3	−36. 3	−37. 2	−38. 2	−39. 1	−40. 1	−41. 0	−42. 0	−42. 9	−43. 8
28.6	−44. 8	−45. 7	−46. 7	−47. 6	−48. 6	−49. 5	−50. 5	−51. 4	−52. 4	−53. 3
28.7	−54. 2	−55. 2	−56. 1	−57. 1	−58. 0	−59. 0	−59. 9	−60. 9	−61. 8	−62. 7
28.8	−63. 7	−64. 6	−65. 6	−66. 5	−67. 5	−68. 4	−69. 4	−70. 3	−71. 2	−72. 2
28.9	−73. 1	−74. 1	−75. 0	−76. 0	−76. 9	−77. 9	−78. 8	−79. 8	−80. 7	−81. 6
29.0	−82. 6	−83. 5	−84. 5	−85. 4	−86. 4	−87. 3	−88. 3	−89. 2	−90. 1	−91. 1
29.1	−92. 0	−93. 0	−93. 9	−94. 9	−95. 8	−96. 7	−97. 7	−98. 6	−99. 6	−100. 5
29.2	−101. 5	−102. 4	−103. 4	−104. 3	−105. 2	−106. 2	−107. 1	−108. 1	−109. 0	−110. 0
29.3	−110. 9	−111. 9	−112. 8	−113. 7	−114. 7	−115. 6	−116. 6	−117. 5	−118. 5	−119. 4
29.4	−120. 3	−121. 3	−122. 2	−123. 2	−124. 1	−125. 1	−126. 0	−127. 0	−127. 9	−128. 8
29.5	−129. 8	−130. 7	−131. 7	−132. 6	−133. 6	−134. 5	−135. 4	−136. 4	−137. 3	−138. 3
29.6	−139. 2	−140. 2	−141. 1	−142. 0	−143. 0	−143. 9	−144. 9	−145. 8	−146. 8	−147. 7
29.7	−148. 6	−149. 6	−150. 5	−151. 5	−152. 4	−153. 4	−154. 3	−155. 3	−156. 2	−157. 1
29.8	−158. 1	−159. 0	−160. 0	−160. 9	−161. 9	−162. 8	−163. 7	−164. 7	−165. 6	−166. 6
29.9	−167. 5	−168. 5	−169. 4	−170. 3	−171. 3	−172. 2	−173. 2	−174. 1	−175. 1	−176. 0
30.0	−176. 9	−177. 9	−178. 8	−179. 8	−180. 7	−181. 6	−182. 6	−183. 5	−184. 5	−185. 4
30.1	−186. 4	−187. 3	−188. 2	−189. 2	−190. 1	−191. 1	−192. 0	−193. 0	−193. 9	−194. 8
30.2	−195. 8	−196. 7	−197. 7	−198. 6	−199. 6	−200. 5	−201. 4	−210. 9	−211. 8	−204. 3
30.3	−205. 2	−206. 1	−207. 1	−208. 0	−209. 0	−209. 9	−210. 9	−211. 8	−212. 7	−213. 7
30.4	−214. 6	−215. 6	−216. 5	−217. 4	−218. 4	−219. 3	−220. 3	−221. 2	−222. 2	−223. 1
30.5	−224. 0	−225. 0	−225. 9	−226. 9	−227. 8	−228. 7	−229. 7	−230. 6	−231. 6	−232. 5

c

*Adopted from Sverdrup, H.U., 1933: Verinfachtes verfahren zum berechnung der druck und massenverteilung im meere, Geophys, Pub. Vol. 10, No. 1, Oslo.

Table II-3a. Conversion of Specific Volume Anomaly to Specific Volume*

$$\alpha = \alpha_{35,0,p} + \delta$$

Example: $T = 4.1°C$
$S = 34.98°/_{00}$
depth $= 1500$ decibars

From

Table IIIb $\delta = 47.5 \times 10^{-5}$
Table IIIa $\alpha_{35,0,p} = 0.96602$
$\alpha = 0.96602 + .000475 = 0.96650$ cm^3/g

Pressure (decibars)	0	100	200	300	400	500	600	700	800	900
0	0.97264	0.97219	0.97174	0.97129	0.97084	0.97040	0.96995	0.96951	0.96907	0.96863
1000	.96819	.96775	.96732	.96688	.96645	.96602	.96559	.96516	.96473	.96430
2000	.96388	.96345	.96303	.96261	.96219	.96177	.96136	.96094	.96053	.96011
3000	.95970	.95929	.95888	.95848	.95807	.95766	.95726	.95686	.95646	.95606
4000	.95566	.95526	.95486	.95447	.95407	.95368	.95329	.95289	.95251	.95212
5000	.95173	.95134	.95096	.95057	.95019	.94981	.94943	.94905	.94867	.94829
6000	.94791	.94754	.94717	.94679	.94642	.94605	.94568	.94531	.94494	.94457
7000	.94421	.94384	.94348	.94312	.94275	.94239	.94203	.94167	.94132	.94096
8000	.94060	.94025	.93989	.93954	.93919	.93883	.93848	.93813	.93778	.93744
9000	.93709	.93674	.93640	.93605	.93571	.93537	.93503	.93469	.93434	.93401

*Adopted from Bjerknes V. and J.W. Sandstrum, 1910: Dynamic Meteorology and Hydrography, Part I: Statics, Carnegie Institution, Pub. 88, Washington, D. C.

Table II-3b Conversion of Thermosteric Anomaly to Specific Volume Anomaly*

$$\delta = \Delta_{s,t} + \delta_{s,p} + \delta_{t,p}$$

For computing the specific volume anomaly of sea water at different temperatures, salinities, and pressure,

Example:
Given
$T = 4.1°C$
$S = 34.98°/_{00}$
depth $= 1500$ decibars
From
Table II-1 $T_t = 27.78$
Table II-2 $\Delta_{st} = 32.8$
Table II-3b $\delta = (32.8 + 0 + 14.7) \times 10^{-5} = 47.5 \times 10^{-5}$.

Table II-3b (continued)

$10^5 \delta_{s,p}$ as a function of salinity and pressure.

Pressure (decibars)	Salinity ‰										
	30	31	32	33	34	35	36	37	38	39	40
0	0.0	0.0	0.0	0.0	0.0	0.0	0.0	0.0	0.0	0.0	0.0
100	−0.8	−0.6	−0.5	−0.3	−0.2	0.0	0.2	0.3	0.4	0.6	0.7
200	−1.5	−1.2	−0.9	−0.6	−0.3	0.0	0.3	0.6	0.9	1.2	1.5
300	−2.3	−1.8	−1.4	−0.9	−0.5	0.0	0.5	0.9	1.3	1.8	2.2
400	−3.0	−2.4	−1.8	−1.2	−0.6	0.0	0.6	1.2	1.8	2.4	3.0
500	−3.8	−3.0	−2.3	−1.5	−0.8	0.0	0.8	1.5	2.2	3.0	3.7
600	−4.5	−3.6	−2.7	−1.8	−0.9	0.0	0.9	1.8	2.6	3.6	4.4
700	−5.3	−4.2	−3.2	−2.1	−1.1	0.0	1.1	2.1	3.1	4.2	5.1
800	−6.0	−4.8	−3.6	−2.4	−1.2	0.0	1.2	2.3	3.5	4.7	5.9
900	−6.8	−5.4	−4.1	−2.7	−1.4	0.0	1.4	2.6	4.0	5.3	6.6
1000	−7.5	−6.0	−4.5	−3.0	−1.5	0.0	1.5	2.9	4.4	5.9	7.3
1100				−3.3	−1.7	0.0	1.7	3.2	4.8	6.5	8.0
1200				−3.6	−1.8	0.0	1.8	3.5	5.2	7.0	8.7
1300				−3.9	−1.9	0.0	1.9	3.8	5.7	7.6	9.4
1400				−4.1	−2.1	0.0	2.1	4.1	6.1	8.1	10.1
1500				−4.4	−2.2	0.0	2.2	4.4	6.5	8.7	10.8
1600				−4.7	−2.3	0.0	2.3	4.7	6.9	9.3	11.5
1700				−4.9	−2.5	0.0	2.5	4.9	7.3	9.8	12.2
1800				−5.2	−2.6	0.0	2.6	5.2	7.8	10.4	12.9
1900				−5.5	−2.7	0.0	2.7	5.5	8.2	10.9	13.6
2000				−5.8	−2.9	0.0	2.9	5.8	8.6	11.5	14.3
2500				−7.2	−3.6	0.0	3.6	7.1	10.7	14.2	17.7
3000				−8.5	−4.3	0.0	4.2	8.5	12.7	16.8	21.0
3500				−9.9	−4.9	0.0	4.9	9.7	14.6	19.4	24.2
4000				−11.1	−5.6	0.0	5.5	11.0	16.5	22.0	27.4

a

Table II-3b (continued)

$10^5 \delta_{s,p}$ as a function of salinity and pressure.

Pressure (decibars)	Salinity ‰				
	34.4	34.6	34.8	35.0	35.2
2000	−1.7	−1.2	−0.6	0.0	0.6
2500	−2.2	−1.4	−0.7	0.0	0.7
3000	−2.6	−1.7	−0.9	0.0	0.8
3500	−2.9	−2.0	−1.0	0.0	1.0
4000	−3.4	−2.2	−1.1	0.0	1.1
4500	−3.7	−2.5	−1.2	0.0	
5000	−4.1	−2.7	−1.4	0.0	
5500	−4.4	−3.0	−1.5	0.0	
6000	−4.8	−3.2	−1.6	0.0	
6500	−5.1	−3.4	−1.7	0.0	
7000	−5.5	−3.6	−1.8	0.0	
7500	−5.8	−3.8	−1.9	0.0	
8000	−6.1	−4.1	−2.0	0.0	
8500	−6.4	−4.3	−2.1	0.0	
9000	−6.7	−4.5	−2.2	0.0	
9500	−7.0	−4.6	−2.3	0.0	
10000	−7.3	−4.8	−2.4	0.0	

b

Table II-3b (continued)

$10^5 \delta_{l,p}$ as a function of temperature and pressure.

Pressure (decibars)	Temperature °C																
	−2	−1	0	1	2	3	4	5	6	7	8	9	10	15	20	25	30
0	0.0	0.0	0.0	0.0	0.0	0.0	0.0	0.0	0.0	0.0	0.0	0.0	0.0	0.0	0.0	0.0	0.0
100	−0.6	−0.3	0.0	0.3	0.5	0.7	1.0	1.2	1.4	1.6	1.8	2.0	2.2	2.9	3.5	3.9	4.2
200	−1.1	−0.6	0.0	0.6	1.1	1.5	2.0	2.4	2.8	3.2	3.5	3.9	4.3	5.8	7.0	7.8	8.4
300	−1.7	−0.9	0.0	0.8	1.6	2.3	3.0	3.7	4.3	4.8	5.4	5.9	6.5	8.8	10.4	11.6	12.5
400	−2.2	−1.1	0.0	1.1	2.1	3.0	4.0	4.9	5.7	6.4	7.2	7.9	8.6	11.7	13.9	15.5	16.7
500	−2.8	−1.4	0.0	1.4	2.6	3.8	5.0	6.1	7.1	8.0	9.0	9.9	10.8	14.6	17.4	19.4	20.9
600	−3.3	−1.7	0.0	1.6	3.1	4.6	6.0	7.3	8.5	9.7	10.8	11.9	12.9	17.4	20.8	23.2	24.9
700	−3.9	−2.0	0.0	1.9	3.7	5.3	6.9	8.4	9.8	11.2	12.5	13.8	15.0	20.2	24.1	27.0	29.0
800	−4.4	−2.2	0.0	2.1	4.2	6.1	7.9	9.6	11.2	12.8	14.3	15.7	17.1	23.1	27.5	30.7	33.0
900	−5.0	−2.5	0.0	2.4	4.7	6.8	8.8	10.7	12.6	14.3	16.0	17.6	19.2	25.9	30.8	34.5	37.1
1000	−5.5	−2.8	0.0	2.7	5.3	7.6	9.8	11.9	14.0	15.9	17.7	19.5	21.3	28.7	34.2	38.3	41.1
1100	−6.0	−3.1	0.0	3.0	5.8	8.3	10.7	13.0	15.3	17.4	19.4	21.4	23.3	31.4	37.5		
1200	−6.5	−3.3	0.0	3.2	6.3	9.1	11.6	14.1	16.6	18.9	21.1	23.3	25.4	34.1	40.7		
1300	−7.1	−3.6	0.0	3.5	6.8	9.8	12.6	15.3	17.9	20.4	22.8	25.1	27.4	36.9	44.0		
1400	−7.6	−3.8	0.0	3.7	7.3	10.6	13.5	16.4	19.2	21.9	24.5	27.0	29.5	39.6	47.2		
1500	−8.1	−4.1	0.0	4.0	7.8	11.3	14.4	17.5	20.5	23.4	26.2	28.9	31.5	42.3	50.5		
1600	−8.6	−4.3	0.0	4.2	8.2	12.0	15.4	18.6	21.8	24.9	27.8	30.7	33.5	44.9	53.6		
1700	−9.1	−4.6	0.0	4.5	8.7	12.7	16.3	19.7	23.1	26.3	29.5	32.5	35.4	47.6	56.8		
1800	−9.6	−4.8	0.0	4.7	9.2	13.4	17.2	20.8	24.3	27.8	31.1	34.3	37.4	50.2	59.9		
1900	−10.1	−5.1	0.0	5.0	9.7	14.1	18.1	21.9	25.6	29.2	32.7	36.1	39.3	52.9	63.1		
2000	−10.6	−5.3	0.0	5.2	10.1	14.7	19.0	23.0	26.9	30.7	34.4	37.9	41.3	55.5	66.2		
2500	−13.1	−6.5	0.0	6.3	12.3	18.1	23.4	28.3	33.1	37.7	42.2	46.6	50.7	68.2			
3000	−15.4	−7.6	0.0	7.4	14.4	21.2	27.6	33.4	39.1	44.6	49.9	55.0	59.8	80.4			
3500	−17.7	−8.7	0.0	8.4	16.4	24.2	31.5	38.3	44.9	51.2	57.3	63.1	68.6	92.2			
4000	−19.9	−9.9	0.0	9.5	18.4	27.0	35.2	43.0	50.5	57.6	64.4	70.9	77.1				

c

Table II-3b (continued)

$10^5 \delta_{t,p}$ as a function of temperature and pressure.

Pressure (decibars)	−1.0	−0.5	0.0	0.5	1.0	1.5	2.0	2.5	3.0	3.5	4.0	4.5	5.0
2000	−5.3	−2.6	0.0	2.6	5.2	7.7	10.1	12.4	14.7	16.9	19.0	21.0	23.0
2500	−6.5	−3.2	0.0	3.2	6.3	9.3	12.3	15.2	18.1	20.8	23.4	25.9	28.3
3000	−7.6	−3.7	0.0	3.7	7.4	11.0	14.4	17.8	21.2	24.5	27.6	30.6	33.4
3500	−8.7	−4.3	0.0	4.2	8.4	12.5	16.4	20.3	24.2	28.0	31.7		
4000	−9.9	−4.9	0.0	4.8	9.5	14.0	18.4	22.7	27.0	31.3	35.6		
4500	−11.0	−5.4	0.0	5.3	10.4	15.4	20.3	25.0	29.6				
5000	−12.0	−5.9	0.0	5.7	11.2	16.7	22.1	27.2	32.1				
5500	−13.0	−6.4	0.0	6.1	12.1	18.0	23.9						
6000	−13.9	−6.9	0.0	6.5	12.9	19.3	25.6						
6500	−14.8	−7.3	0.0	6.9	13.7	20.5	27.3						
7000	−15.7	−7.8	0.0	7.4	14.6	21.7	28.9						
7500	−16.6	−8.2	0.0	7.8	15.4	23.0	30.5						
8000	−17.4	−8.6	0.0	8.2	16.1	24.1	32.0						
8500	−18.2	−9.0	0.0	8.6	17.0	25.3	33.5						
9000	−18.9	−9.3	0.0	9.0	17.8	26.4	34.9						
9500	−19.6	−9.7	0.0	9.4	18.6	27.5	36.2						
10000	−20.3	−10.0	0.0	9.8	19.4	28.6	37.5						

Temperature °C

d

*Adopted from Sverdrup, H.U., 1933: Verinfachtes verfahren zum berechnung der druck und massenverteilung im meere, Geophys, Pub. Vol. 10, No. 1, Oslo.

Table II-4 Compressibility of Distilled Water and Seawater*

Compressibility k is defined as $k = 1/\rho \ \partial\rho/\partial\rho$ under isothermal conditions. In table pressure is given in bars (1 bar = 10^6 dynes/cm²). Example: for distilled water at a pressure of 100 bars and a temperature of 10°C, the compressibility is 46.6×10^{-6} bars $^{-1}$.

Compressibility of Distilled Water ($\times 10^6$ bars)

	T (°C)				
P (bars)	0	10	20	30	40
1	50.7	47.9	46.1	44.9	44.4
100	49.2	46.6	44.9	43.8	43.3
200	47.9	45.4	43.8	42.7	42.3
300	46.5	44.2	42.6	41.7	41.3
400	45.3	43.1	41.6	40.7	40.3
500	44.1	42.0	40.6	39.7	39.4
600	42.9	40.9	39.6	38.8	38.5
700	41.8	39.9	38.7	37.9	37.6
800	40.7	39.0	37.8	37.0	36.8
900	39.7	38.0	36.9	36.2	35.9
1000	38.7	37.1	36.0	35.4	35.2

a

Compressibility of Seawater

10⁰/₀₀ S

	T (°C)				
P (bars)	0	10	20	30	40
1	49.5	46.9	45.1	44.0	43.5
100	48.1	45.6	43.9	42.9	42.4
200	46.8	44.4	42.8	41.8	41.4
300	45.5	43.3	41.8	40.8	40.4
400	44.3	42.2	40.7	39.9	39.5
500	43.1	41.1	39.7	38.9	38.6
600	41.9	40.0	38.8	38.0	37.7
700	40.8	39.1	37.8	37.1	36.8
800	39.8	38.1	37.0	36.3	36.0
900	38.8	37.2	36.1	35.5	35.2
1000	37.8	36.3	35.3	34.7	34.4

b

Table II-4 (continued)

$20^0/_{00}$ S

P (bars)	T (°C)				
	0	10	20	30	40
1	48.4	45.8	44.1	43.1	42.6
100	47.0	44.6	43.0	42.0	41.6
200	45.7	43.4	41.9	41.0	40.6
300	44.4	42.3	40.9	40.0	39.6
400	43.2	41.2	39.8	39.0	38.7
500	42.1	40.2	38.9	38.1	37.8
600	41.0	39.2	37.9	37.2	36.9
700	39.9	38.2	37.0	36.4	36.1
800	38.9	37.3	36.2	35.5	35.3
900	37.9	36.4	35.3	34.7	34.5
1000	37.0	35.5	34.5	33.9	33.7

c

$30^0/_{00}$ S

P (bars)	T (°C)				
	0	10	20	30	40
1	47.2	44.8	43.2	42.1	41.7
100	45.9	43.6	42.0	41.1	40.7
200	44.6	42.5	41.0	40.1	39.7
300	43.4	41.4	40.0	39.1	38.8
400	42.2	40.3	39.0	38.2	37.8
500	41.1	39.3	38.0	37.3	37.0
600	40.0	38.3	37.1	36.4	36.1
700	39.0	37.3	36.2	35.6	35.3
800	38.0	36.4	35.4	34.8	34.5
900	37.0	35.6	34.6	34.0	33.8
1000	36.1	34.7	33.8	33.2	33.0

d

Table II-4 (continued)

35⁰/₀₀ S

	T (°C)				
P (bars)	0	10	20	30	40
1	46.7	44.3	42.7	41.7	41.2
100	45.4	43.1	41.6	40.6	40.2
200	44.1	42.0	40.5	39.6	39.3
300	42.9	40.9	39.5	38.7	38.3
400	41.7	39.8	38.5	37.8	37.4
500	40.6	38.8	37.6	36.9	36.6
600	39.5	37.8	36.7	36.0	35.7
700	38.5	36.9	35.8	35.2	34.9
800	37.5	36.0	35.0	34.4	34.1
900	36.6	35.1	34.2	33.6	33.4
1000	35.6	34.3	33.4	32.8	32.6

e

40⁰/₀₀ S

	T (°C)				
P (bars)	0	10	20	30	40
1	46.1	43.8	42.2	41.2	40.8
100	44.8	42.6	41.1	40.2	39.8
200	43.6	41.5	40.1	39.2	38.8
300	42.4	40.4	39.1	38.3	37.9
400	41.2	39.4	38.1	37.3	37.0
500	40.1	38.4	37.2	36.5	36.2
600	39.1	37.4	36.3	35.6	35.3
700	38.0	36.5	35.4	34.8	34.5
800	37.1	35.6	34.6	34.0	33.8
900	36.1	34.7	33.8	33.2	33.0
1000	35.2	33.9	33.0	32.5	32.3

f

*From Wilson, W. and D. Bradley, 1966: Specific volume, thermal expansion and isothermal compressibility of seawater, Technical Report NOLTR 66-103, U.S. Naval Ordnance Laboratory, White Oak, Maryland.

Table II-5 Thermal Expansion of Distilled Water and Seawater*

Thermal expansion c is defined as $c = 1/\rho\ \partial\rho/\partial T$. In table pressure is given in bars (1 bar = 10^6 dynes/cm²).

Example: for distilled water at a pressure of 100 bars and a temperature of 10°C the thermal expansion is 11.2×10^{-5} per degree Celsius.

Coefficient of Thermal Expansion† of Distilled water ($\times 10^5$ °C)

	T (°C)				
P (bars)	0	10	20	30	40
1	−3.4	9.0	20.0	30.3	40.7
100	−0.2	11.2	21.4	31.1	41.0
200	2.8	13.3	22.8	31.9	41.2
300	5.6	15.2	24.0	32.6	41.4
400	8.3	17.1	25.2	33.2	41.5
500	10.8	18.9	26.4	33.8	41.7
600	13.2	20.6	27.5	34.4	41.8
700	15.5	22.1	28.5	35.0	41.9
800	17.7	23.6	29.5	35.5	42.0
900	19.7	25.0	30.4	36.0	42.1
1000	21.6	26.4	31.2	36.4	42.2

a

Coefficient of Thermal Expansion

10⁰/₀₀ S

	T (°C)				
P (bars)	0	10	20	30	40
1	−0.2	11.1	21.2	30.7	40.4
100	2.8	13.2	22.5	31.5	40.6
200	5.7	15.2	23.8	32.2	40.8
3000	8.5	17.1	25.1	32.9	41.0
400	11.1	18.9	26.2	33.5	41.2
500	13.5	20.6	27.3	34.1	41.3
600	15.8	22.2	28.4	34.7	41.4
700	18.0	23.7	29.4	35.2	41.5
800	20.1	25.2	30.3	35.7	41.6
900	22.0	26.5	31.2	36.1	41.7
1000	23.9	27.8	32.0	36.6	41.7

b

Table II-5 (continued)

20⁰/₀₀ S

			T (°C)		
P (bars)	0	10	20	30	40
1	3.0	13.2	22.3	31.1	40.1
100	6.0	15.2	23.6	31.9	40.3
2000	8.7	17.1	24.9	32.6	40.5
300	11.4	19.0	26.1	33.2	40.7
400	13.8	20.7	27.2	33.8	40.8
500	16.2	22.3	28.3	34.4	40.9
600	18.4	23.8	29.2	34.9	41.0
700	20.5	25.3	30.2	35.4	41.1
800	22.5	26.7	31.1	35.9	41.2
900	24.4	28.0	31.9	36.3	41.3
1000	26.1	29.2	32.7	36.7	41.3

c

30⁰/₀₀ S

			T (°C)		
P (bars)	0	10	20	30	40
1	6.2	15.3	23.5	31.5	39.8
100	9.0	17.2	24.8	32.2	40.0
200	11.7	19.1	26.0	32.9	40.2
300	14.2	20.8	27.1	33.5	40.3
400	16.6	22.5	28.2	34.1	40.4
500	18.8	24.0	29.2	34.6	40.5
600	21.0	25.5	30.1	35.1	40.6
700	23.0	26.9	31.0	35.6	40.7
800	24.9	28.2	31.9	36.0	40.8
900	26.6	29.4	32.7	36.4	40.9
1000	28.3	30.6	33.4	36.8	40.9

d

Table II-5 (continued)

35⁰/₀₀ S

P (bars)	T (°C)				
	0	10	20	30	40
1	7.8	16.3	24.1	31.7	39.6
100	10.6	18.2	25.3	32.4	39.8
200	13.2	20.0	26.5	33.0	40.0
300	15.6	21.7	27.6	33.6	40.1
400	18.0	23.3	28.6	34.2	40.2
500	20.2	24.9	29.6	34.7	40.3
600	22.2	26.3	30.6	35.2	40.4
700	24.2	27.7	31.4	35.7	40.5
800	26.0	29.0	32.3	36.1	40.6
900	27.8	30.2	33.1	36.5	40.6
1000	29.4	31.3	33.8	36.9	40.7

e

40⁰/₀₀ S

P (bars)	T (°C)				
	0	10	20	30	40
1	9.4	17.3	24.7	31.9	39.5
100	12.1	19.2	25.9	32.6	39.6
200	14.7	21.0	27.0	33.2	39.8
300	17.1	22.6	28.1	33.8	39.9
400	19.4	24.2	29.1	34.3	40.1
500	21.5	25.7	30.1	34.8	40.2
600	23.5	27.1	31.0	35.3	40.2
700	25.4	28.5	31.9	35.8	40.3
800	27.2	29.7	32.7	36.2	40.4
900	29.0	30.9	33.4	36.6	40.4
1000	30.6	32.0	34.2	37.0	40.5

f

*From Wilson, W. and D. Bradley, 1966: Specific volume, thermal expansion and isothermal compressibility of seawater, Technical Report NOLTR 66-103, U.S. Naval Ordnance Laboratory, White Oak, Maryland.

Table II-6 Adiabatic Cooling of Water Raised to the Surface*

Example: a water particle at 5000 meters with a temperature of $+1.0°$ Celsius and a salinity of 34.85%/00 if raised to the surface under adiabatic conditions would be cooled 0.434° and have a temperature of 0.57° Celsius. More recent work by Cox and Smith (Table II-6b) suggests the earlier values of Helland-Hansen are slightly low.

Adiabatic cooling (in 0.01°C) in seawater (S %/00 = 34.85) of a temperature in situ T_m(°C) if this water is raised from a depth d to the surface (according to Helland-Hansen).

T_m (°C) d (m)	-2	-1	0	1	2	3	4	5	6	7	8	9	10
1,000	2.6	3.5	4.4	5.3	6.2	7.0	7.8	8.6	9.5	10.2	11.0	11.7	12.4
1,500	5.4	6.7	8.0	9.3	10.6	11.8	12.9	14.1	15.3	16.4	17.5	18.6	19.7
2,000	7.2	8.9	10.7	12.4	14.1	15.7	17.2	18.8	20.4	21.9	23.3	24.8	26.2
2,500	10.9	12.9	15.1	17.2	19.3	21.2	23.1	25.0	26.9	28.7	30.6	32.3	34.0
3,000	13.6	16.1	18.7	21.2	23.6	25.9	28.2	30.5	32.7	34.9	37.1	39.2	41.2
3,500	18.1	21.0	24.0	26.8	29.6	32.2	34.8	37.3	39.8	42.3	44.8	47.2	49.5
4,000	21.7	25.0	28.4	31.6	34.7	37.7	40.6	43.5	46.3	49.1	51.9	54.6	57.2
4,500	27.0	30.7	34.4	37.9	41.3	44.7	47.9						
5,000	31.5	35.5	39.6	43.4	47.2	50.9	54.4						
5,500	37.5	41.9	46.3	50.4	54.5	58.4	62.2						
6,000	42.8	47.5	52.2	56.7	61.1	65.3	69.4						
6,500			59.6	64.3	69.0	73.4	77.7						
7,000			66.2	71.3	76.2	80.9	85.5						
7,500			74.2	79.5	84.7	89.6	94.4						
8,000			81.5	87.1	92.5	97.7	102.7						
8,500			90.1	95.9	101.6	106.9	112.1						
9,000			98.1	104.1	109.9	115.6	121.0						
9,500			107.2	113.4	119.4	125.2	130.8						
10,000			115.7	122.1	128.3	134.4	140.2						

a

Table II-6 (continued)

Adiabatic cooling (in 0.01°C) in seawater (S $^0/_{00}$ = 34.85) of a temperature in situ T_m (°C) if this water is raised from a depth d to the surface, according to Cox and Smith (1959). Results of Helland-Hansen (1930) are given in parentheses.

T_m(°C) d (m)	−2	0	5	10	15	20
1000	2.7(2.6)	4.6(4.4)	9.2(8.6)	13.0(12.4)	16.7(16.1)	20.0(19.5)
2000	7.5(7.2)	11.4(10.7)	19.9(18.8)	27.4(26.2)	—	—
3000	14.4(13.6)	20.1(18.7)	32.6(30.5)	—	—	—
4000	23.2(21.7)	30.8(28.4)	46.7(43.5)	—	—	—
6000	47.2(42.8)	57.8(52.2)	80.2(73.6)	—	—	—
8000	—	91.0(81.5)	119.2(107.6)	—	—	—

b

Adiabatic cooling (in 0.01°C) when seawater of different salinity ($S^0/_{00}$) and temperature T_m(°C) is raised from 1000-m depth to the surface (according to Heland-Hansen).

T_m(°C) $S^0/_{00}$	0	2	4	6	8	10	12	14	16	18	20	22
30.0	3.5	5.3	7.0	8.7	10.3	11.8	13.2	14.7	16.1	17.6	18.9	20.3
32.0	3.9	5.7	7.3	9.0	10.6	12.1	13.5	15.0	16.4	17.8	19.1	20.5
34.0	4.3	6.0	7.7	9.4	10.9	12.4	13.8	15.3	16.6	18.0	19.3	20.7
36.0	4.7	6.4	8.1	9.7	11.2	12.7	14.1	15.5	16.9	18.3	19.6	20.9
38.0	5.1	6.8	8.4	10.0	11.6	13.0	14.4	15.8	17.2	18.5	19.8	21.1

c

*After Helend-Hansen, B., 1930: Physical Oceanography and Meteorology, "Michael Sars" North Atlantic Deep Sea Expedition, 1910, Rept. Sci. Results, Vol. 1, Art 2; and Cox, R.A. and N.D. Smith 1959: The specific heat of seawater, Proc. Roy. Soc. A, Vol. 252, London as adapted by Neuman, G. and W.J. Pierson, Jr., 1966: Principles of Physical Oceanography, Prentice-Hall, Englewood Cliffs, New Jersey.

Table II-7 Heat Capacity of Seawater at 1 Atm (cal/g °C)*

Temperature (°C)	S (⁰/oo)				
	0	10	20	30	40
0	1.005	0.989	0.974	0.959	0.946
10	1.002	0.987	0.973	0.960	0.947
20	1.000	0.986	0.973	0.961	0.949
30	0.999	0.986	0.974	0.962	0.950
40	0.998	0.986	0.974	0.963	0.951

*From Bromlay, L.A., V.A. Desaussure, J.C. Clipp, and J.S. Wright, 1967: J. Chem. Eng. Data, Vol. 12, as adapted by Horne, R.A. 1969: Marine Chemistry, John Wiley and Sons, New York.

Table II-8 Molecular Viscosity of Seawater*

Salinity† (⁰/oo)	Temperature (°C)						
	0	5	10	15	20	25	30
0........	17.9	15.2	13.1	11.4	10.1	8.9	8.0
10........	18.2	15.5	13.4	11.7	10.3	9.1	8.2
20........	18.5	15.8	13.6	11.9	10.5	9.3	8.4
30........	18.8	16.0	13.8	12.1	10.7	9.5	8.6
35........	18.9	16.1	13.9	12.2	10.9	9.6	8.7

*From Dorsey, N.E. 1940: Properties of ordinary water substance, Amer. Chem. Soc. Monograph Ser. No. 81, Reinhold Pub. Corp., New York.

†Molecular viscosity of distilled water and seawater at atmospheric pressure. Units of molecular viscosity, μ, are in 10^{-3} gm/cm s. The viscosity of seawater at 5°C and 35⁰/oo is .0161 gm/cm s.

Table II-9 Velocity of Sound in Seawater*

$C = 1449.22 + \Delta C_t + \Delta C_p + \Delta C_s + \Delta C_{s,t,p}$ in m/s

Example: depth = 1000 decibars

$T = 4.0°C$

$S = 35.00°/oo$

According to tables (a) through (d),

$C = 1449.22 + 17.64 + 16.16 - 0.00 - 0.10$

$= 1483$ m/s

T	0.0	0.1	0.2	0.3	0.4	0.5	0.6	0.7	0.8	0.9
					ΔC_t					
−3	−14.37	−14.87	−15.36	−15.86	−16.36	−16.86	−17.36	−17.87	−18.37	−18.88
−2	−9.47	−9.95	−10.44	−10.93	−11.41	−11.90	−12.39	−12.89	−13.38	−13.87
−1	−4.68	−5.15	−5.63	−6.10	−6.58	−7.06	−7.54	−8.02	−8.50	−8.98
0	0.00	−0.46	−0.93	−1.39	−1.86	−2.33	−2.79	−3.36	−3.73	−4.21
0	0.00	0.46	0.92	1.38	1.84	2.30	2.75	3.21	3.66	4.12
1	4.57	5.02	5.47	5.92	6.37	6.81	7.26	7.70	8.15	8.59
2	9.03	9.47	9.91	10.35	10.79	11.22	11.66	12.09	12.52	12.96
3	13.39	13.82	14.24	14.67	15.10	15.53	15.95	16.37	16.80	17.22
4	17.64	18.06	18.48	18.89	19.31	19.73	20.14	20.55	20.97	21.38
5	21.79	22.20	22.60	23.01	23.42	23.82	24.23	24.63	25.03	25.43
6	25.84	26.23	26.63	27.03	27.43	27.82	28.22	28.61	29.00	29.39
7	29.78	30.17	30.56	30.95	31.34	31.72	32.11	32.49	32.87	33.25
8	33.64	34.02	34.39	34.77	35.15	35.53	35.90	36.27	36.65	37.02
9	37.39	37.76	38.13	38.50	38.87	39.23	39.60	39.96	40.33	40.69
10	41.05	41.41	41.77	42.13	42.49	42.85	43.20	43.56	43.91	44.27
11	44.62	44.97	45.32	45.67	46.02	46.37	46.72	47.06	47.41	47.75
12	48.10	48.44	48.78	49.12	49.46	49.80	50.14	50.48	50.81	51.15
13	51.48	51.82	52.15	52.48	52.81	53.14	53.47	53.80	54.13	54.46
14	54.78	55.11	55.43	55.76	56.08	56.40	56.72	57.04	57.36	57.68
15	58.00	58.31	58.63	58.94	59.26	59.57	59.88	60.19	60.50	60.81
16	61.12	61.43	61.74	62.04	62.35	62.65	62.96	63.26	63.56	63.87
17	64.17	64.47	64.76	65.06	65.36	65.66	65.95	66.25	66.54	66.83
18	67.13	67.42	67.71	68.00	68.29	68.58	68.87	69.15	69.44	69.72
19	70.01	70.29	70.57	70.86	71.14	71.42	71.70	71.98	72.26	72.53
20	72.81	73.09	73.36	73.63	73.91	74.18	74.45	74.72	74.99	75.26
21	75.53	75.80	76.07	76.34	76.60	76.87	77.13	77.39	77.66	77.92
22	78.18	78.44	78.70	78.96	79.22	79.48	79.73	79.99	80.24	80.50
23	80.75	81.01	81.26	81.51	81.76	82.01	82.26	82.51	82.76	83.01
24	83.25	83.50	83.74	83.99	84.23	84.47	84.72	84.96	85.20	85.44
25	85.68	85.92	86.16	86.39	86.63	86.87	87.10	87.34	87.57	87.80
26	88.04	88.27	88.50	88.73	88.96	89.19	89.42	89.64	89.87	90.10
27	90.32	90.55	90.77	91.00	91.22	91.44	91.66	91.88	92.10	92.32
28	92.54	92.76	92.98	93.19	93.41	93.63	93.84	94.06	94.27	94.48
29	94.69	94.91	95.12	95.33	95.54	95.75	95.96	96.16	96.37	96.58
30	96.78	96.99	97.19	97.40	97.60	97.80	98.00	98.21	98.41	98.61

(a)

Table II-9 (continued)

$$\Delta C_p$$

P†	0	10	20	30	40	50	60	70	80	90
0	—	1.61	3.21	4.82	6.44	8.05	9.67	11.29	12.91	14.53
100	16.16	17.79	19.42	21.05	22.68	24.32	25.96	27.60	29.24	30.89
200	32.54	34.19	35.84	37.50	39.15	40.81	42.47	44.14	45.81	47.47
300	49.15	50.82	52.49	54.17	55.85	57.54	59.22	60.91	62.60	64.29
400	65.98	67.68	69.38	71.08	72.78	74.49	76.19	77.90	79.61	81.33
500	83.04	84.76	85.48	88.20	89.92	91.65	93.38	95.10	96.84	98.57
600	100.30	102.04	103.78	105.52	107.26	109.00	110.75	112.49	114.24	115.99
700	117.74	119.49	121.25	123.00	124.76	125.52	128.28	130.04	131.80	133.56
800	135.33	137.09	138.86	140.62	142.39	144.16	145.93	147.70	149.47	151.24
900	153.01	154.78	156.56	158.33	160.10	161.88	163.65	165.42	167.20	168.97

†Pressure in bars.

(b)

$$\Delta C_s$$

S	0.0	0.1	0.2	0.3	0.4	0.5	0.6	0.7	0.8	0.9
33	−3.09	−2.93	−2.76	−2.59	−2.43	−2.26	−2.10	−1.94	−1.78	−1.62
34	−1.47	−1.32	−1.16	−1.01	−0.86	−0.72	−0.57	−0.42	−0.28	−0.14
35	0.00	0.14	0.28	0.41	0.54	0.68	0.81	0.94	1.06	1.19
36	1.31	1.44	1.56	1.68	1.79	1.91	2.03	2.14	2.26	2.36
37	2.47	2.58	2.68	2.79	2.89	2.99	3.09	3.19	3.28	3.38

(c)

Table II-9 (continued)

$$\Delta C_{s,t,p}$$

P	S										T									
		-4	-2	0	2	4	6	8	10	12	14	16	18	20	22	24	26	28	30
1.0332	33	-0.09	-0.05	0.00	0.05	0.09	0.14	0.19	0.24	0.29	0.33	0.38	0.43	0.48	0.53	0.57	0.62	0.67	0.72
	34	-0.05	-0.02	0.00	0.02	0.05	0.07	0.09	0.12	0.14	0.17	0.19	0.21	0.24	0.26	0.29	0.31	0.34	0.36
	35	0.00	0.00	0.00	0.00	0.00	0.00	0.00	0.00	0.00	0.00	0.00	0.00	0.00	0.00	0.00	0.00	0.00	0.00
	36	0.05	0.02	0.00	-0.02	-0.05	-0.07	-0.10	-0.12	-0.15	-0.17	-0.19	-0.22	-0.22	-0.26	-0.26	-0.31	-0.33	-0.36
	37	0.10	0.05	0.00	-0.05	-0.10	-0.14	-0.19	-0.24	-0.29	-0.34	-0.38	-0.43	-0.48	-0.53	-0.57	-0.62	-0.67	-0.72
100	33	0.00	-0.03	-0.05	-0.06	-0.05	-0.04	-0.02	0.02	0.06	0.11	0.17	0.24	0.31	0.39	0.48	0.58	0.68	0.79
	34	0.07	0.02	-0.02	-0.06	-0.08	-0.09	-0.09	-0.08	-0.06	-0.04	-0.00	-0.04	0.09	0.15	0.22	0.29	0.37	0.45
	35	0.14	0.07	0.00	-0.06	-0.10	-0.14	-0.16	-0.18	-0.18	-0.18	-0.17	-0.15	-0.13	-0.09	-0.05	-0.01	0.05	0.11
	36	0.22	0.11	0.02	-0.06	-0.13	-0.19	-0.24	-0.28	-0.31	-0.33	-0.34	-0.35	-0.35	-0.34	-0.32	-0.30	-0.27	-0.23
	37	0.29	0.16	0.05	-0.06	-0.15	-0.23	-0.31	-0.37	-0.43	-0.48	-0.51	-0.54	-0.57	-0.58	-0.59	-0.59	-0.58	-0.57
200	33	0.12	0.01	-0.09	-0.16	-0.21	-0.25	-0.26	-0.25	-0.23	-0.19	-0.13	-0.05	0.04	0.15	0.27	0.41	0.56	0.72
	34	0.22	0.08	-0.04	-0.14	-0.22	-0.28	-0.31	-0.33	-0.33	-0.32	-0.28	-0.23	-0.16	-0.08	0.02	0.13	0.25	0.39
	35	0.31	0.14	0.00	-0.12	-0.23	-0.31	-0.37	-0.41	-0.44	-0.44	-0.43	-0.41	-0.37	-0.31	-0.24	-0.15	-0.05	0.06
	36	0.41	0.21	0.04	-0.10	-0.23	-0.34	-0.42	-0.49	-0.54	-0.57	-0.59	-0.59	-0.57	-0.54	-0.49	-0.43	-0.35	-0.26
	37	0.50	0.28	0.09	-0.08	-0.24	-0.37	-0.48	-0.57	-0.65	-0.70	-0.74	-0.76	-0.77	-0.76	-0.74	-0.70	-0.65	-0.59
300	33	0.29	0.07	-0.12	-0.27	-0.40	-0.49	-0.55	-0.59	-0.59	-0.57	-0.52	-0.45	-0.35	-0.23	-0.09	0.07	0.26	0.46
	34	0.40	0.15	-0.06	-0.24	-0.39	-0.50	-0.59	-0.65	-0.68	-0.68	-0.66	-0.61	-0.54	-0.45	-0.33	-0.19	-0.03	0.14
	35	0.51	0.24	0.00	-0.20	-0.38	-0.52	-0.63	-0.71	-0.77	-0.80	-0.80	-0.78	-0.73	-0.66	-0.57	-0.46	-0.33	-0.18
	36	0.62	0.32	0.06	-0.17	-0.37	-0.53	-0.67	-0.78	-0.86	-0.91	-0.94	-0.94	-0.92	-0.88	-0.81	-0.73	-0.62	-0.49
	37	0.73	0.41	0.12	-0.13	-0.36	-0.55	-0.71	-0.84	-0.95	-1.03	-1.08	-1.11	-1.12	-1.10	-1.06	-0.99	-0.91	-0.81
400	33	0.49	0.15	-0.15	-0.40	-0.60	-0.77	-0.89	-0.98	-1.03	-1.05	-1.03	-0.97	-0.88	-0.77	-0.62	-0.44	-0.24	-0.01
	34	0.61	0.25	-0.97	-0.35	-0.58	-0.77	-0.92	-1.04	-1.11	-1.15	-1.16	-1.13	-1.07	-0.98	-0.85	-0.70	-0.53	-0.32
	35	0.73	0.34	0.00	-0.30	-0.56	-0.78	-0.95	-1.09	-1.19	-1.26	-1.29	-1.29	-1.25	-1.18	-1.09	-0.96	-0.81	-0.63
	36	0.86	0.44	0.07	-0.25	-0.54	-0.78	-0.98	-1.15	-1.27	-1.37	-1.42	-1.44	-1.43	-1.39	-1.32	-1.22	-1.10	-0.95
	37	0.98	0.54	0.15	-0.21	-0.52	-0.78	-1.01	-1.20	-1.36	-1.47	-1.55	-1.60	-1.62	-1.60	-1.56	-1.48	-1.38	-1.26

500	33	0.73	0.26	−0.16	−0.53	−0.84	−1.09	−1.30	−1.46	−1.57	−1.64	−1.66	−1.64	−1.58	−1.48	−1.35	−1.17	−0.97	−0.73
	34	0.86	0.36	−0.08	−0.47	−0.81	−1.09	−1.32	−1.51	−1.65	−1.74	−1.79	−1.80	−1.76	−1.69	−1.58	−1.43	−1.25	−1.04
	35	1.00	0.47	0.00	−0.42	−0.78	−1.09	−1.35	−1.56	−1.72	−1.84	−1.92	−1.95	−1.94	−1.89	−1.81	−1.69	−1.53	−1.34
	36	1.13	0.58	0.08	−0.36	−0.75	−1.08	−1.37	−1.61	−1.80	−1.94	−2.04	−2.10	−2.12	−2.10	−2.04	−1.94	−1.82	−1.65
	37	1.26	0.68	0.16	−0.31	−0.72	−1.08	−1.39	−1.66	−1.87	−2.04	−2.17	−2.26	−2.30	−2.30	−2.27	−2.20	−2.10	−1.96
600	33	1.02	0.39	−0.17	−0.67	−1.10	−1.47	−1.78	−2.03	−2.22	−2.36	−2.45	−2.48	−2.47	−2.40	−2.30	−2.14	−1.95	−1.72
	34	1.16	0.50	−0.09	−0.61	−1.07	−1.47	−1.80	−2.08	−2.30	−2.46	−2.57	−2.63	−2.65	−2.61	−2.53	−2.40	−2.24	−2.03
	35	1.30	0.62	0.00	−0.55	−1.04	−1.46	−1.82	−2.12	−2.37	−2.56	−2.70	−2.79	−2.82	−2.81	−2.76	−2.66	−2.52	−2.34
	36	1.44	0.73	0.09	−0.49	−1.00	−1.45	−1.84	−2.17	−2.44	−2.66	−2.83	−2.94	−3.00	−3.02	−2.99	−2.92	−2.80	−2.65
	37	1.58	0.84	0.17	−0.43	−0.97	−1.45	−1.86	−2.22	−2.52	−2.76	−2.95	−3.09	−3.18	−3.22	−3.22	−3.18	−3.09	−2.96
700	33	1.37	0.56	−0.17	−0.83	−1.40	−1.91	−2.34	−2.70	−3.00	−3.23	−3.40	−3.51	−3.56	−3.55	−3.49	−3.38	−3.22	−3.02
	34	1.51	0.67	−0.09	−0.77	−1.37	−1.90	−2.36	−2.75	−3.07	−3.33	−3.53	−3.66	−3.74	−3.76	−3.73	−3.64	−3.51	−3.33
	35	1.65	0.79	0.00	−0.71	−1.34	−1.90	−2.38	−2.80	−3.15	−3.43	−3.65	−3.82	−3.92	−3.97	−3.96	−3.91	−3.80	−3.65
	36	1.79	0.90	0.09	−0.65	−1.31	−1.89	−2.40	−2.85	−3.22	−3.53	−3.78	−3.97	−4.10	−4.18	−4.20	−4.17	−4.09	−3.97
	37	1.93	1.01	0.17	−0.59	−1.27	−1.88	−2.42	−2.89	−3.30	−3.64	−3.91	−4.13	−4.29	−4.39	−4.44	−4.43	−4.38	−4.28
800	33	1.77	0.76	−0.17	−1.00	−1.74	−2.41	−2.98	−3.48	−3.91	−4.25	−4.53	−4.73	−4.87	−4.95	−4.96	−4.91	−4.81	−4.65
	34	1.91	0.87	−0.08	−0.94	−1.72	−2.40	−3.01	−3.54	−3.99	−4.36	−4.66	−4.90	−5.06	−5.17	−5.21	−5.19	−5.11	−4.98
	35	2.05	0.98	0.00	−0.89	−1.69	−2.40	−3.04	−3.59	−4.07	−4.47	−4.80	−5.06	−5.25	−5.38	−5.45	−5.46	−5.41	−5.31
	36	2.19	1.09	0.08	−0.83	−1.66	−2.40	−3.06	−3.64	−4.15	−4.58	−4.93	−5.22	−5.44	−5.60	−5.69	−5.73	−5.71	−5.63
	37	2.32	1.20	0.17	−0.78	−1.63	−2.40	−3.09	−3.70	−4.23	−4.69	−5.07	−5.39	−5.63	−5.82	−5.94	−6.00	−6.01	−5.96
900	33	2.24	0.99	−0.15	−1.19	−2.13	−2.98	−3.73	−4.39	−4.96	−5.45	−5.86	−6.19	−6.44	−6.62	−6.73	−6.77	−6.75	−6.66
	34	2.37	1.09	−0.08	−1.14	−2.11	−2.98	−3.76	−4.45	−5.05	−5.57	−6.00	−6.36	−6.64	−6.85	−6.98	−7.05	−7.06	−7.00
	35	2.50	1.20	0.00	−1.09	−2.09	−2.99	−3.80	−4.51	−5.14	−5.69	−6.15	−6.53	−6.84	−7.08	−7.24	−7.34	−7.37	−7.34
	36	2.63	1.30	0.08	−1.05	−2.07	−3.00	−3.83	−4.58	−5.23	−5.81	−6.30	−6.71	−7.04	−7.31	−7.50	−7.62	−7.68	−7.68
	37	2.76	1.40	0.15	−1.00	−2.05	−3.01	−3.87	−4.64	−5.32	−5.92	−6.44	−6.88	−7.24	−7.53	−7.75	−7.91	−7.99	−8.02
1000	33	2.76	1.26	−0.13	−1.40	−2.57	−3.62	−4.57	−5.42	−6.18	−6.83	−7.40	−7.88	−8.27	−8.58	−8.81	−8.97	−9.06	−9.07
	34	2.88	1.35	−0.06	−1.37	−2.56	−3.64	−4.62	−5.50	−6.28	−6.97	−7.56	−8.07	−8.49	−8.83	−9.09	−9.27	−9.39	−9.43
	35	3.01	1.44	0.00	−1.33	−2.55	−3.66	−4.67	−5.58	−6.39	−7.10	−7.72	−8.26	−8.71	−9.07	−9.36	−9.57	−9.71	−9.79
	36	3.13	1.54	0.06	−1.29	−2.54	−3.68	−4.72	−5.65	−6.49	−7.23	−7.88	−8.45	−8.92	−9.32	−9.63	−9.88	−10.04	−10.14
	37	3.25	1.63	0.13	−1.26	−2.53	−3.70	−4.77	−5.73	−6.59	−7.36	−8.04	−8.63	−9.14	−9.56	−9.91	−10.18	−10.37	−10.50

(d)

*After Wilson, W.D., 1960: Equation for the speed of sound in seawater, Journ. of Acoustical Soc. Amer., Vol. 32.

Table II-10 Maximum Density of Seawater, Temperature of Maximum Density, and Freezing Point; All As a Function of Salinity*

$S(^0/oo)$	$\sigma_{t\ max}$	T_{max} (°C)	T_f (°C)	$S(^0/oo)$	$\sigma_{t\ max}$	T_{max} (°C)	T_f† (°C)
0	0	3.947	−0.014	21	16.87	−0.529	−1.137
1	0.85	3.743	−0.066	22	17.67	−0.744	−1.192
2	1.69	3.546	−0.118	23	18.48	−0.964	−1.248
3	2.51	3.347	−0.170	24	19.29	−1.180	−1.303
4	3.33	3.133	−0.223	25	20.10	−1.398	−1.359
5	4.15	2.926	−0.275	26	20.91	−1.613	−1.414
6	4.96	2.713	−0.328	27	21.72	−1.831	−1.470
7	5.77	2.501	−0.381	28	22.53	−2.048	−1.526
8	6.58	2.292	−0.434	29	23.34	−2.262	−1.582
9	7.38	2.075	−0.487	30	24.15	−2.473	−1.638
10	8.18	1.860	−0.541	31	24.97	−2.687	−1.695
11	8.97	1.645	−0.594	32	25.78	−2.900	−1.751
12	9.76	1.426	−0.648	33	26.59	−3.109	−1.808
13	10.56	1.210	−0.702	34	27.40	−3.318	−1.865
14	11.35	0.994	−0.758	35	28.22	−3.524	−1.922
15	12.13	0.772	−0.810	36	29.04	−3.733	−1.979
16	12.92	0.562	−0.864	37	29.86	−3.936	−2.036
17	13.69	0.342	−0.918	38	30.68	−4.138	−2.094
18	14.48	0.124	−0.973	39	31.50	−4.340	−2.151
19	15.27	−0.090	−1.028	40	32.32	−4.541	−2.209
20	16.07	−0.310	−1.082	41	33.14	−4.738	−2.267

*After Dorsey, N.E., 1940: Properties of ordinary water substance, Amer. Chem. Soc., Monograph Ser. No. 81, Reinhold Pub. Corp., New York; and Doherty, B.T. and D.R. Kester, 1974: Freezing point of seawater, J. Mar. Res., Vol. 32.

Table II-11 Temperature of Crystalization*

	T(°C)
Pure ice (H$_2$0)	−1.9
Salt	
Sodium sulfate (Na$_2$SO$_4$ · 10H$_2$O)	−8.2
Sodium chloride	
(NaCl, probably as hydrate)	−23
Magnesium chloride (MgCl$_2$ · 12H$_2$O)	−36
Potassium chloride (KCl)	−36
Calcium chloride (CaCl$_2$)	−55

*After Ringer, W.E., 1928: On the composition of seawater fog during freezing, Conseil Perm, Intern, p. l'Exp de lar Mem Rapp et Proc.

Table II-12 Light Absorption* of Typical Seawater†

Sample	Wavelength (nm)							
	360	400	500	520	600	700	750	800
Pure water	0.001	0.001	0.002	0.002	0.010	0.025	0.115	0.086
Artificial seawater	0.011	0.003	0.005	0.007	0.010	0.025	0.115	0.086
Ocean water, unfiltered	0.012	0.009	0.007	0.008	0.011	0.025	0.115	0.086
Continental slope waters, unfiltered	0.052	0.030	0.011	0.010	0.012	0.035	0.130	0.088
Continental slope waters, filtered	0.016	0.010	0.005	0.005	0.012	0.030	0.115	0.086
Inshore water, unfiltered	0.055	0.042	0.028	0.026	0.035	0.052	0.140	0.100
Inshore water, filtered	0.015	0.010	0.005	0.005	0.010	0.025	0.110	0.086

*Absorption for 10 cm path.

†From Clarke, G.L. and H.R. James, 1939: J. Opt. Soc. Am., Vol. 29.

Table II-13 Miscellaneous Constants and Conversions

Earth

Equatorial Radius	= 6378 km
Polar Radius	= 6356 km
Mass	= 5.976×10^{27} gm
Volume	= 1.083×10^{12} km^3
Area	= 362×10^6 km^3
Density	= 5.52 gm/cm^3
Seconds in year	= 31,560,000
Angular Velocity	= 7.29×10^{-5} per sec
Solar Constant	= 1.94 ± .03 cal/cm^2 sec
	= 0.135 ± .002 watts/cm^2

Conversion

1 nautical mile = 6076 ft = 1.852 $\cong$ 1 min of latitude

1 knot = 1 nautical mile/hr = 1.85 km/hr = 51.4 cm/sec = 1.15 stat mile/hr

1 fathom = 6 ft = 1.83 m

1 bar = 10^6 dynes/cm^2 = 0.987 atmosphere

1° latitude = 111 km

INDEX

A

Absolute vorticity, 289
Absorption:
  electromagnetic waves in water, 250
  solar radiation in air, 30-32, 56, 57
  solar radiation in water, 43, 44
  sound waves in air, 250
  sound waves in water, 246, 249, 250
  visible spectrum in water, 262-65
    tables, 330
Acceleration: 74-76
  derivation, 272, 273
  field acceleration, 76
  local acceleration, 75, 273
  particle acceleration, 75, 273
Adiabatic change in temperature, 7, 8
  tables, 322-23
Albedo, 32, 33, 56, 57
Ambient noise, 250-52
Antarctic Bottom Water, 168-73

Antarctic Circumpolar Current, 136, 138,
  160, 162-65
  effect of bottom topography, 162
Antarctic, polar front, 164, 165
Attenuation: (*see also* Absorption)
  sound energy in water, 243, 249, 250
  visible spectrum in water, 243, 263-67

B

Back radiation (*see* Long wave radiation)
Beers' law, 249, 263
Benguela Current, 138, 150
Bottom water formation, 171, 172
Bowen's radio, 40
Box models, 68-71
Brazil Current, 137-38
Breaking waves, 216, 217
  internal waves, 188, 229
Brunt-Väisälä frequency, 18, 229

C

California Current, 150
Capillary wave (*see* Surface tension
    waves)
Change in sea level:
    caused by geostrophic currents, 241
    caused by thermal expansion, 240
Climatic change, 57-59
Conservation equation, 61-67
    derivation, 269-73, 281-82
Conservation of mass (*see* Mass balance)
Conservation of salt (*see* Salt balance)
Conservative property, 63
Continuity equation, 62, 64
    derivation, 269, 270
Coordinate system for earth, 72
Core analysis (*see* Water masses analysis)
Coriolis acceleration/force, 78-84
    centrifugal acceleration, 81
    derivation, 276-80
    effect on gravity (*see* Eötövös correc-
        tion)
    latitudinal effect, 133, 134
    near equator, 156, 158, 159
Cromwell Current, 156-58, (*see also*
    Equatorial Currents)
Current direction, definition, 73
Current measurements:
    Antarctic, 162, 165
    California Current, 151
    Cromwell Current, 157
    deep North Atlantic, 191
    deep Western boundary currents, 173,
        174
    Gulf Stream, 141, 143, 144, 147
    Indian Ocean, 161
    Kuroshio, 149
    tropics, 154, 155

D

Decibar, 92
Decibels, 244, 245
Deep sea trenches, 8, 184, 185
Deep sound channel, 256-58

Deep water waves (*see* Surface gravity
    waves)
Degree-day, 51, 52
Density, seawater, 10-15
    sigma-t, 10, 12-15, 166
    sigma-t tables, 297-307
    sigma-θ, 10, 12-14, 17, 166, 167
Density distribution (*figures*):
    Gulf Stream, 142
    Oceanwide, 167
Diffusion, 65-67, (*see also* Molecular dif-
    fusion and Eddy diffusion)
Dispersive waves, 202-6, 215, (*see also*
    Group velocity)
Doldrums (*see* Intertropical convergence)
Double diffusion, 185-87, 189
Drag coefficient, 108
Dynamic height, 13, 95, 96

E

Echo sounding, 259-61
Eddy diffusion, 66, 67, 167, 168, 287, 288
    coefficient, 67
    mesoscale turbulence, 188-91
Eddy viscosity, 66, 87, 121, 280-83
    coefficient, 88
Edge Waves, 217, 226
Ekman circulation, 121-25, 128, 192
    spiral, 121, 122
    transport, 121, 122, 128
Eötövös correction, 85, 278
Equation of motion, 74, 89, 279
Equation of state of seawater, 12
Equatorial Currents, 118, 138, 151-61
    Countercurrent, 118, 137, 138, 152,
        154, 158
Estuaries, 97-106
    conservation equation, 99
    convective flow, 97-100
    effect of Coriolis force, 104
    effect of wind stress, 104, 105
    net flow vs tidal flow, 104, 106
    tidal mixing, 103
    type, 100-104
Evaporation, 29, 34, 37-39, 56-58, 60, 69

F

Foucault pendulum, 79-80
Friction, 85-89, 97, 129-34, (*see also* Viscosity, Reynolds stress)
  proportional to velocity, 88, 129
  surface gravity waves, 213, 214
Fully developed sea, 210-12

G

Geostrophic currents, 112-19, 174
  level of no motion, 114
  slope of sea surface, 113, 114, 116-19
  slope of thermocline, 114, 116-19
Geostrophic equation, 112
Geostrophic motion, effect of friction, 129-31
Gravity, 74, 84, 85, 90, (*see also* Eötövös correction)
Group velocity: (*see also* Dispersive waves)
  derivation, 294
  surface gravity waves, 203, 204, 215, 216, 220
  surface tension waves, 206
Gulf Stream, 38, 53, 114-16, 135-47
  meanders, 53, 140, 144, 145
  rings, 145-47, 188, 189
  speed, 140, 141, 143, 145
  transport, 145, 147

H

Halocline, 25, 26
Heat balance, 29, 42-58
  ice, 49-52
  global, 52-56
  radiation, 36, 54
Heat flow from earth, 28
Heat transfer: (*see also* Longwave radiation, Evaporation)
  advection, 43, 52, 53, 57
  ocean-atmosphere, 29, 40, 41, 45-47, 49, 50, 55-57
Hydrostatic equation, 92

I

Inclined plane:
  effect of Coriolis force, 82-83
  effect of Coriolis force and friction, 129-31
Inertial motion, 126, 128, 129
Inertial period, 80
Intermediate waves (*see* Surface gravity waves)
Internal waves, 227-29
  velocity, 227, 228
  effect on mixing, 185, 188
Intertropical convergence, 151
Inverse barometer effect, 226, 240
Irradiance, 263, 264

K

Kuroshio, 53, 135, 137-39, 145, 148, 149, 150

L

Labrador Current, 137, 138
Langmuir circulation, 192-93
Light (*see* Solar radiation)
Longshore currents, 217, 219
Longwave radiation, 28, 29, 33, 34, 56, 57

M

Margule's equation, 117-20
Mass balance, 61-63, 68, 99
Mass transport (*see* Transport)
Mediterranean outflow water, 168, 169, 171
Microstructure, 184-88, 250
Mixed layer, 4, 47-49, 193
Mixing time, 68-71
Molecular:
  diffusion, 65, 66, 286, 287
  thermal conductivity, 66
  viscosity, 66, 85, 108, 249
    decay of kinetic energy, 87
    tables, 324

Monsoon, 124, 126, 137, 160
  effect on surface temperature, 124, 127

N

Neutrally bouyant floats, 189, 191
Newton's first law, 78, 126
Newton's second law, 73, 74
Nonconservative property, 63
North equatorial current (*see* Equatorial
  Currents)
Norwegian overflow water, 169-73

O

Ocean statistics:
  area, 2
  depth, 3, 4
  salinity, 182
  specific volume anomaly, 182
  temperature, 182
  volume, 2
Oxygen:
  depletion rate, 63
  distribution in tropics, 155-57
  trenches, 184
Oyashio, 137, 138

P

Pacific Common Water, 172
Partial differential equations, 64
Peru Current, 150, 151
Planetary vorticity, 133-35
Potential energy, 16, 130
Potential temperature, 7, 8, 13
  tables, 322-23
Potential vorticity, 290
Precipitation, 42, 58, 60, 69
Pressure gradient force, 76, 77
  derivation, 274-76
  slope of interface, 93, 94, 95
  slope of sea surface, 77, 93
Pycnocline, 26

R

Raleigh scattering, 31, 32, 263
Refraction:
  light, 262, 263
  sound, 253-58
    critical angle, 253, 254
    surface gravity waves, 215-18
Relative vorticity, 289
Residence time, 71
  derivation, 296
Reynolds number, 108
Reynolds stress, 88, 89, 288
  derivation, 283-86
Rip currents, 217, 219
River flow, 60, 96, 97

S

Salinity:
  constituent elements, 11
  definition, 8, 9, 62
  measurement, 9, 60
  range, 9, 10, 182
Salinity distribution:
  Antarctica, 163
  deep water, Atlantic, 170-72
  Mediterranean outflow, 168, 169
  tropical Pacific, 155
Salt balance, 60, 62-64, 68, 69, 99
Salt fingers (*see* Double diffusion)
Scattering:
  solar radiation, 30-32, 56
  sound, 249, 250
  visible spectrum, 262-66
Sea ice, 19-27, 49-52, 171
  freezing point, 329
  temperature of crystalization, 27, 330
Seawater, color, 266
Seawater, properties, 18-25
  compressibility, 19-21, 24
    tables, 316-18
  electrical conductivity, 20, 23
  freezing point, 19, 23
    tables, 329
  heat capacity, 19, 20
    tables, 324

Seawater, properties (*continued*):
  maximum density, 23, 24
    tables, 329
  thermal expansion, 20, 22-24, 240
    tables, 319-21
Seiche, 222-26
Sensible heat conduction (*see* Heat transfer)
Settling velocity (*see* Terminal velocity)
Shallow water waves (*see* Surface gravity waves)
Sigma-*t* (*see* Density)
Sigma-*θ* (*see* Density)
Slope of sea surface: (*see* Pressure gradient force)
  at equator, 158
SOFAR floats (*see* Neutrally bouyant floats)
Solar constant, 30, 331
Solar radiation, 28-33, 35
  absorption (*see* Absorption)
  penetration in seawater, 43, 44, 47, 243, 262
  reflection (*see* Albedo)
  scattering (*see* Scattering, Raleigh scattering)
  spectral distribution, 30, 31
Sonar equation, 246-51
Sound intensity, 244, 245
Sound level, 244-51
Sound, properties of:
  absorption (*see* Absorption)
  attenuation (*see* Attenuation)
  scattering (*see* Scattering)
  energy flux, 244, 246
  frequency, 246, 250, 259, 262
  pressure, 244-46, 248
  reference, 244, 245
Sound rays, 253-58
Sound reflection, 258
Sound, target definition, 259-62, (*see also* Target strength)
Sound transmission, 246-50, 256-58
  cylindrical spreading, 247-49
  spherical spreading, 246-48
Sound velocity, 17, 19, 25, 245, 246, 253-58
  tables, 325-28

Snell's law, 253, 254, 262, 263
Specific volume:
  range, 182
  tables, 311
  specific volume anomaly, 12, 13, 15, 96
    tables, 312-15
  thermosteric anomaly, 12, 13, 15
    tables, 308-10
South equatorial current (*see* Equatorial Currents)
Stability, 8, 13, 15-18, 47, 67, 186
Standing wave, 222, 223
Steady state, 42, 64
Stefan-Boltzman law, 33
Stokes law, 108
Storm surges, 226, 227
Surface gravity waves:
  particle motion, 199-201, 222-24
  pressure, 201, 202
  significant wave height, 210
  speed, 196-99
    derivation, 291-94
Surface slicks, 207
Surface tension waves, 194, 205-7

T

Target strength, 248-49
Temperature:
  in situ, 7
  potential, 7, 8
  range, 182
Temperature and salinity distribution:
  bivariate histograms, 178, 180, 181
  temperature, salinity diagrams, 174-79
Temperature distribution (*figures*):
  Antarctic, 163, 164
  Arabian Sea surface, 127
  deep Atlantic, 183
  deep Pacific, 173
  deep trenches, 184
  Eastern Pacific surface, 125
  Gulf Stream, 116, 142
  Pacific, 5
  Peru Current, 153
  tropical Pacific, 118, 155
Terminal velocity, 106-10

Thermocline, 5, 6, 26, 230, 253, (*see also* Temperature distribution)
  diurnal, 49, 50
  effect on stability, 154
  formation, 43-49
  seasonal, 6, 43-48
  spreading of isotherms, 151, 156, 157
  tropical, 118, 154-56
Thermohaline circulation, 166-71
Tidal components, 239, 240
Tidal currents, 241, 242
Tidal force, 229-33
Tidal height, 236
Tidal prediction, 238-42
Tides, 229-42
  amphidromic point, 235
  Bay of Fundy, 234, 242
  equilibrium, 233
  dynamic, 234
  particle velocity, 241
  resonance, 234, 235
  rotary, 235, 237, 238
  shallow water wave, 234
Torque, 131, 132
Transmittance, 264, 265
Transport:
  currents:
    Antarctic, 162
    California, 150
    deep Pacific, 174
    Gulf Stream, 145, 147
  definition, 68
Tritium, half-life, 63
Tsunami, 220-22
Turbulence, 37, 38, 66, 67, 87, 167, 188-91, (*see also* Eddy diffusion, Eddy viscosity, **Reynolds** stress)

U

Underwater sound (*see* Sound)
Upwelling, 44, 122-28, 174
  biological productivity, 122, 123, 151
  surface temperature, 124, 125, 127, 153

V

Vertical velocity, 193
Viscosity (*see* Eddy viscosity, Molecular viscosity)
Volume transport (*see* Transport)
Vorticity, 131-35
  equation derivation, 289-91

W

Water balance, 58, 60, (*see also* Mass balance)
Water masses:
  analysis, 166-72, 174-84
  characteristics, 174-82
  formation, 171, 172
Wave equation, 196-98
  derivation, 291-94
Wave generation, 205-10, 212
  fetch, 210, 212
Wave propagation, 212-17
Wave refraction (*see* Refraction)
Waves: (*see also* Surface gravity waves, Surface tension waves, Internal waves)
  amplitude, 195, 196, 201, 214
  energy, 194, 202-5, 210, 211, 215-17, 222
  energy flux, 203-5, 216, 221
  frequency, 194, 195, 196, 211
  group velocity (*see* Group velocity)
  height, 195, 196, 201, 202, 211, 213, 214
  length, 195, 196
  number, 195, 196
  particle motion, 196, 241
  period, 194-96, 211
Wave spectrum, 195, 208-13
Western boundary currents, 131-35, 139
  deep, 172-74
Wien's law, 33
Wind direction, definition, 73
Wind speed, 38, 85, 120, 126, 160, 193
Wind stress, 85, 86, 88, 120, 132-34, 158

Table of Symbols

a	attenuation coefficient of sound	$\mathbf{i}$	unit vector in x direction
a	wave amplitude	I	sound intensity
A	coefficient of eddy diffusion, viscosity	j	coefficient of molecular diffusion
		$\mathbf{j}$	unit vector in y direction
b	attenuation coefficient of sound	J	frictional constant
B	Bowen's ratio	k	coefficient of compressibility
B	area	k	coefficient of thermal conductivity of water
c	undefined constant		
c	speed of sound	k_i	coefficient of thermal conductivity of ice
c_p	specific heat of water		
c_f	heat of fusion	$\mathbf{k}$	unit vector in z direction
C	phase velocity of wave	K	Kelvin temperature
D	dynamic height	l	length, length of basin
e_a	vapor pressure of air at *in situ* relative humidity and air temperature	L	sound level
		m	mass
e_w	vapor pressure of air at 100% relative humidity and surface temperature of water	M	mass of moon
		M	mass transport per unit length
		N	Brunt-Väisälä frequency
E	energy	N	noise level (dB)
E	stability	O	oxygen concentration
f	frequency	p	pressure
f	Coriolis parameter, $2\,\Omega$ s in ϕ	P	distance of earth from moon
F	force	q	R cos ϕ
F	flux	Q	heat flux (Q_s sun, Q_b effective back radiation, Q_e evaporation, Q_h sensible, Q_T heat gain)
g	acceleration due to gravity		
G	gravitational constant		
h	depth of ocean	r	radius
H	wave height	r_i	radius of inertial circle
H	tritium	r_c	radius of curvature
i	slope of sea surface, or density interface	R	radius of earth
		$\mathbf{R}$	distance or range